# Use R!

*Series Editors:*
Robert Gentleman    Kurt Hornik    Giovanni Parmigiani

For further volumes:
http://www.springer.com/series/6991

# Use R!

*Albert:* Bayesian Computation with R
*Bivand/Pebesma/Gómez-Rubio:* Applied Spatial Data Analysis with R
*Cook/Swayne:* Interactive and Dynamic Graphics for Data Analysis:
 With R and GGobi
*Hahne/Huber/Gentleman/Falcon:* Bioconductor Case Studies
*Paradis:* Analysis of Phylogenetics and Evolution with R
*Pfaff:* Analysis of Integrated and Cointegrated Time Series with R
*Sarkar:* Lattice: Multivariate Data Visualization with R
*Spector:* Data Manipulation with R

Radhakrishnan Nagarajan • Marco Scutari
Sophie Lèbre

# Bayesian Networks in R

## with Applications in Systems Biology

Radhakrishnan Nagarajan
Division of Biomedical Informatics
Department of Biostatistics
University of Kentucky
Lexington, Kentucky, USA

Marco Scutari
Genetics Institute
University College London
London, United Kingdom

Sophie Lèbre
ICube
Université de Strasbourg
France

ISBN 978-1-4614-6445-7     ISBN 978-1-4614-6446-4 (eBook)
DOI 10.1007/978-1-4614-6446-4
Springer New York Heidelberg Dordrecht London

Library of Congress Control Number: 2013935127

Printed on acid-free paper

Springer is part of Springer Science+Business Media (www.springer.com)

*To Adriana Brogini and Fortunato Pesarin,*
*who showed me what an academic should be.*

# Preface

*Real world entities work in concert as a system and not in isolation. Understanding the associations between these entities from their digital signatures can provide novel system-level insights and is an important step prior to developing meaningful interventions.*

While there have been significant advances in capturing data from the entities across complex real-world systems, their associations and relationships are largely unknown. Associations between the entities may reveal interesting system-level properties that may not be apparent otherwise. Often these associations are hypothesized by superimposing knowledge across distinct reductionist representations of these entities obtained from disparate sources. Such representations, while useful, may provide only an incomplete picture of the associations. This can be attributed to their dependence on prior knowledge and failure of the principle of superposition in general. Such representations may also be unhelpful in discovering novel undocumented associations. A more rigorous approach would be to identify associations from data measured simultaneously across the entities of interest from a given system. These data sets or digital signatures are quantized in time and amplitude and in turn may (dynamic) or may not (static) contain explicit temporal information. Symmetric measures such as correlation have been helpful in modeling direct associations as undirected graphs. However, it is well appreciated that the association between a given pair of entities may be indirect and often mediated through others. Symmetric measures are also immune to the direction of association by their very definition. Graphical models such as Bayesian networks have especially proven to be useful in this regard. The vertices (nodes) represent the entities of interest, the arcs (edges) represent their associations, and the entire Bayesian network represents the joint probability distribution between the entities of interest. Bayesian networks may also reveal possible causal relationships between these entities under certain implicit assumptions. More specifically, their ability to model associations from observational data sets where no active perturbation is possible has drawn attention across a wide spectrum of disciplines including biology, medicine, and health care.

There have been several noteworthy contributions to Bayesian network modeling and inference along with open-source implementations of the related algorithms. However, many of these prior contributions are extremely involved and demand a high level of sophistication from the reader. This book is unique as it introduces the reader to the essential concepts in conjunction with examples in the open-source statistical environment R. The level of sophistication is gradually increased across the chapters. Each chapter is accompanied by examples and exercises with solutions for enhanced understanding and experimentation. Thus this book may appeal to multidisciplinary audience and can potentially assist in teaching graduate-level courses in Bayesian networks and inference that permit hands-on experimentation of the concepts and approaches. The data sets considered essentially consist of publicly available molecular expression profiles. The emphasis on molecular data can be attributed to the growing need in life sciences for discovering novel associations across biological paradigms with minimal precedence and increasing emphasis on data-driven approaches. Classical studies in life sciences have focused on understanding the changes in the expression of a given set of molecules, such as genes and proteins, across distinct phenotypes and disease states. However, with recent advances in high-throughput assays that enable simultaneous screening of a large number of genes, there has been growing interest in understanding the associations between these molecules that may provide system-level insights. Such system-level insights have been argued to be critical prior to developing meaningful interventions. These efforts together fall under the emerging discipline called systems biology. Bayesian networks have especially proven to be useful abstractions of the underlying biological pathways and signaling mechanisms. Their usefulness is also exemplified by their ability to discover new associations in addition to validating known associations between the entities of interest.

While a list of popular open-source R packages pertinent to Bayesian networks is listed under Table 2.1 (Chap. 2), the discussion focuses on the packages **bnlearn**, **G1DBN**, and **ARTIVA**.

> http://cran.r-project.org/web/packages/bnlearn
> http://cran.r-project.org/web/packages/G1DBN
> http://cran.r-project.org/web/packages/ARTIVA

We believe that these packages are comprehensive and accommodate the necessary functionalities required across the chapters. We also believe that concentrating on these packages keeps the book more focused with minimal demand on the audience time in learning the functionalities across the various open-source R packages.

This book is organized as follows. Chapter 1 introduces the reader to the essentials of graph theory and R programming. Chapter 2 discusses the essential definitions and properties of Bayesian networks with an emphasis on static Bayesian networks. It introduces the reader to structure and parameter learning from multiple independent realizations of data sets without explicit temporal information. Such data sets are quite common and represent a snapshot of the process. The impact of discretization on the network inference with application to molecular expression data is also discussed. The lack of temporal information implicitly excludes the presence of feedback or cycles, resulting in a directed acyclic graphical representation of the associations between the entities. These limitations are overcome by learning networks from data sets with explicit temporal signatures. In Chap. 3, we discuss the usefulness of dynamic Bayesian networks for learning the network structure in the presence of explicit temporal information such as multivariate time series. Homogeneous and nonhomogeneous dynamic Bayesian networks are discussed. In Chap. 4, static and dynamic Bayesian network inference methods are discussed. Some of the network learning algorithms discussed in the earlier chapters are computationally intensive limiting their usefulness across large and high-dimensional data sets. Parallelization options for some of the algorithms discussed in the earlier chapters are discussed in Chap. 5 to overcome some of these limitations.

Lexington, KY                                                         Radhakrishnan Nagarajan
London, UK                                                                       Marco Scutari
Strasbourg, France                                                              Sophie Lèbre

# Contents

**1 Introduction** ........................................................ 1
  1.1 A Brief Introduction to Graph Theory ......................... 1
    1.1.1 Graphs, Nodes, and Arcs ............................... 1
    1.1.2 The Structure of a Graph .............................. 2
    1.1.3 Further Reading ...................................... 4
  1.2 The R Environment for Statistical Computing .................. 4
    1.2.1 Base Distribution and Contributed Packages ............. 4
    1.2.2 A Quick Introduction to R ............................. 5
    1.2.3 Further Reading ...................................... 10
  Exercises ....................................................... 11

**2 Bayesian Networks in the Absence of Temporal Information** ........ 13
  2.1 Bayesian Networks: Essential Definitions and Properties ......... 13
    2.1.1 Graph Structure and Probability Factorization ........... 13
    2.1.2 Fundamental Connections .............................. 15
    2.1.3 Equivalent Structures ................................. 15
    2.1.4 Markov Blankets ..................................... 16
  2.2 Static Bayesian Networks Modeling ........................... 17
    2.2.1 Constraint-Based Structure Learning Algorithms ......... 17
    2.2.2 Score-Based Structure Learning Algorithms ............. 19
    2.2.3 Hybrid Structure Learning Algorithms ................. 20
    2.2.4 Choosing Distributions, Conditional Independence
        Tests, and Network Scores ............................ 20
    2.2.5 Parameter Learning .................................. 23
    2.2.6 Discretization ....................................... 23
  2.3 Static Bayesian Networks Modeling with R .................... 24
    2.3.1 Popular R Packages for Bayesian Network Modeling ...... 24
    2.3.2 Creating and Manipulating Network Structures .......... 26
    2.3.3 Plotting Network Structures .......................... 34
    2.3.4 Structure Learning .................................. 35

2.3.5   Parameter Learning ................................... 40
2.3.6   Discretization ....................................... 42
2.4   Pearl's Causality ............................................ 44
2.5   Applications to Gene Expression Profiles...................... 46
2.5.1   Model Averaging .................................... 47
2.5.2   Choosing the Significance Threshold ................... 51
2.5.3   Handling Interventional Data ......................... 53
Exercises ....................................................... 56

3   **Bayesian Networks in the Presence of Temporal Information** ........ 59
3.1   Time Series and Vector Auto-Regressive Processes ............. 59
3.1.1   Univariate Time Series............................... 59
3.1.2   Multivariate Time Series ............................. 60
3.2   Dynamic Bayesian Networks: Essential Definitions and Properties . 63
3.2.1   Definitions........................................... 63
3.2.2   Dynamic Bayesian Network Representation
of a VAR Process ................................... 66
3.3   Dynamic Bayesian Network Learning Algorithms ............... 67
3.3.1   Least Absolute Shrinkage and Selection Operator ........ 67
3.3.2   James–Stein Shrinkage .............................. 68
3.3.3   First-Order Conditional Dependencies Approximation ..... 68
3.3.4   Modular Networks ................................... 69
3.4   Non-homogeneous Dynamic Bayesian Network Learning ........ 69
3.5   Dynamic Bayesian Network Learning with R .................. 72
3.5.1   Multivariate Time Series Analysis ..................... 72
3.5.2   LASSO Learning: **lars** and **simone** ..................... 74
3.5.3   Other Shrinkage Approaches: **GeneNet**, **G1DBN** ......... 78
3.5.4   Non-homogeneous Dynamic Bayesian Network
Learning: **ARTIVA** ................................. 80
Exercises ....................................................... 81

4   **Bayesian Network Inference Algorithms** ......................... 85
4.1   Reasoning Under Uncertainty ............................... 85
4.1.1   Probabilistic Reasoning and Evidence ................... 85
4.1.2   Algorithms for Belief Updating: Exact and Approximate
Inference .......................................... 87
4.1.3   Causal Inference ..................................... 90
4.2   Inference in Static Bayesian Networks ........................ 91
4.2.1   Exact Inference ...................................... 91
4.2.2   Approximate Inference ............................... 93
4.3   Inference in Dynamic Bayesian Networks ..................... 94
Exercises ...................................................... 100

**5   Parallel Computing for Bayesian Networks** ...................... 103
      5.1   Foundations of Parallel Computing ........................... 103
      5.2   Parallel Programming in R ................................... 105
      5.3   Applications to Structure and Parameter Learning .............. 108
            5.3.1   Constraint-Based Structure Learning Algorithms ......... 109
            5.3.2   Score-Based Structure Learning Algorithms ............. 112
            5.3.3   Hybrid Structure Learning Algorithms .................. 114
            5.3.4   Parameter Learning ................................. 115
      5.4   Applications to Inference Procedures ......................... 115
            5.4.1   Bootstrap .......................................... 115
            5.4.2   Cross-Validation .................................... 117
            5.4.3   Conditional Probability Queries ....................... 120
      Exercises ..................................................... 123

**Solutions** ....................................................... 125

**References** ...................................................... 149

**Index** .......................................................... 155

# Chapter 1
# Introduction

**Abstract** Bayesian networks and their applications to real-world problems lie at the intersection of several fields such as probability and graph theory. In this chapter a brief introduction to the terminology and the basic properties of graphs, with particular attention to directed graphs, is provided. As with other Use R!-series books, a brief introduction to the R environment and basic R programming is also provided. Some background in probability theory and programming is assumed. However, the necessary references are included under the respective sections for a more complete treatment.

## 1.1 A Brief Introduction to Graph Theory

### 1.1.1 Graphs, Nodes, and Arcs

A graph $G = (\mathbf{V}, A)$ consists of a nonempty set $\mathbf{V}$ of *nodes* or *vertices* and a finite (but possibly empty) set $A$ of pairs of vertices called *arcs*, *links*, or *edges*.

Each arc $a = (u, v)$ can be defined either as an ordered or an unordered pair of nodes, which are said to be *connected* by and *incident* on the arc and to be *adjacent* to each other. Since they are adjacent, $u$ and $v$ are also said to be *neighbors*. If $(u, v)$ is an ordered pair, $u$ is said to be the *tail* of the arc and $v$ the *head*; then the arc is said to be *directed* from $u$ to $v$ and is usually represented with an arrowhead in $v$ ($u \rightarrow v$). It is also said that the arc *leaves* or is *outgoing* for $u$ and that it *enters* or is *incoming* for $v$. If $(u, v)$ is unordered, $u$ and $v$ are simply said to be incident on the arc without any further distinction. In this case, they are commonly referred to as *undirected arcs* or *edges*, denoted with $e \in E$ and represented with a simple line ($u - v$).

The characterization of arcs as directed or undirected induces an equivalent characterization of the graphs themselves, which are said to be *directed graphs* (denoted with $G = (\mathbf{V}, A)$) if all arcs are directed, *undirected graphs* (denoted with

R. Nagarajan et al., *Bayesian Networks in R: with Applications in Systems Biology*, Use R! 48, DOI 10.1007/978-1-4614-6446-4_1,
© Springer Science+Business Media New York 2013

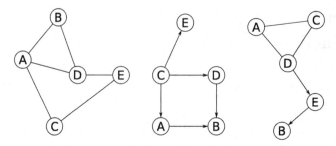

**Fig. 1.1** An undirected graph (*left*), a directed graph (*center*) and a partially directed graph (*right*)

$G = (\mathbf{V}, E)$) if all arcs are undirected, and *partially directed* or *mixed graphs* (denoted with $G = (\mathbf{V}, A, E)$) if they contain both directed and undirected arcs.

Examples of *directed*, *undirected*, and mixed *partially directed* graphs are shown in Fig. 1.1 in that order. For the undirected graph, Fig. 1.1:

- The node set is $\mathbf{V} = \{A, B, C, D, E\}$ and the edge set is $E = \{ (A - B), (A - C), (A - D), (B - D), (C - E), (D - E) \}$.
- Arcs are undirected, so, i.e., $A - B$ and $B - A$ are equivalent and identify the same edge.
- Likewise, A is connected to B, B is connected to A, and A and B are adjacent.

For the directed graph, Fig 1.1:

- The node set is $\mathbf{V} = \{A, B, C, D, E\}$ and the graph is characterized by an arc set $A = \{(A \rightarrow B), (C \rightarrow A), (D \rightarrow B), (C \rightarrow D), (C \rightarrow E)\}$ instead of an edge set $E$.
- Arcs are directed, so, i.e., $A \rightarrow B$ and $B \rightarrow A$ identify different arcs. For instance, $A \rightarrow B \in A$ while $B \rightarrow A \notin A$. Under the additional constraint of acyclicity, it is not possible for both arcs to be present in the graph because there can be at most one arc between each pair of nodes.
- Also, A and B are adjacent, as there is an arc $(A \rightarrow B)$ from A to B. $A \rightarrow B$ is an outgoing arc for A (the tail), an incoming arc for B (the head), and an incident arc for both A and B.

On the other hand, the partially directed graph, Fig. 1.1, is characterized by the combination of an edge set $E = \{(A - C), (A - D), (C - D)\}$ and an arc set $A = \{(D \rightarrow E), (E \rightarrow B)\}$.

An undirected graph can always be constructed from a directed or partially directed one by substituting all the directed arcs with undirected ones; such a graph is called the *skeleton* or the *underlying undirected graph* of the original graph.

### 1.1.2 The Structure of a Graph

The pattern with which the arcs appear in a graph is referred to as either the *structure* of the graph or the *configuration* of the arcs. In the context of this book it is assumed

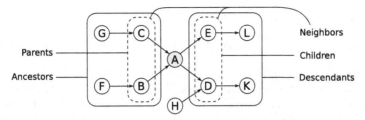

**Fig. 1.2** Parents, children, ancestors, descendants, and neighbors of node A in a directed graph

that the vertices $u$ and $v$ incident on each arc are distinct and that there is at most one arc between them so that $(u,v)$ uniquely identifies an arc. This definition also implicitly excludes presence of a *loop* that can occur when $u = v$.

The simplest structure is an *empty graph*, i.e., a graph with no arcs. On the other end of the spectrum are *saturated graphs*, in which each node is connected to every other node. Real-world graphical abstractions usually fall between these two extremes and can be either *sparse* or *dense*. While the distinction between these two classes of graphs is rather vague, a graph is usually considered sparse if $O(|E| + |A|) = O(|\mathbf{V}|)$.

The structure of a graph can reveal interesting statistical properties. Some of the most important ones deal with *paths*. Paths are essentially sequences of arcs or edges *connecting* two nodes, called *end-vertices* or *end-nodes*. Paths are denoted with the sequence of vertices $(v_1, v_2, \ldots, v_n)$ incident on those arcs. The arcs connecting the vertices $v_1, v_2, \ldots, v_n$ are assumed to be unique, so that a path passes through each arc only once. In directed graphs it is also assumed that all the arcs in a path follow the same direction, and we say that a path *leads from* $v_1$ (i.e., the tail of the first arc in the path) *to* $v_n$ (i.e., the head of the last arc in the path). In undirected and mixed graphs (and in general when referring to a graph regardless which class it belongs to), arcs in a path can point in either direction or be undirected. Paths in which $v_1 = v_n$ are called *cycles* and are treated with particular care in Bayesian network theory.

The structure of a directed graph defines a partial ordering of the nodes if the graph is *acyclic*, that is, if it does not contain any cycle or loop. This ordering is called an *acyclic* or *topological ordering* and is induced by the direction of the arcs. It is defined as follows: if a node $v_i$ precedes $v_j$, there can be no arc from $v_j$ to $v_i$. According to this definition the first nodes are the *root nodes*, which have no incoming arcs, and the last ones are the *leaf nodes*, which have at least one incoming arc but no outgoing ones. Furthermore, if there is a path leading from $v_i$ to $v_j$, $v_i$ precedes $v_j$ in the sequence of the ordered nodes. In this case $v_i$ is called an *ancestor* of $v_j$ and $v_j$ is called a *descendant* of $v_i$. If the path is composed by a single arc, by analogy $x_i$ is a *parent* of $v_j$ and $v_j$ is a *child* of $v_i$.

Consider, for instance, node A in the directed acyclic graph shown in Fig. 1.2. Its neighborhood is the union of the parents and children; adjacent nodes necessarily fall into one of these two categories. Its parents are also ancestors, as they necessarily precede $A$ in the topological ordering. Likewise, children are also descendants.

The topological ordering induced by the graph structure is

$$(\{F,G,H\},\{C,B\},\{A\},\{D,E\},\{L,K\}). \tag{1.1}$$

The nodes are only *partially ordered*; for example, no ordering can be established among root nodes or leaf nodes. As a result, in practice the topological ordering of a directed acyclic graph is defined over a set of unordered sets of nodes, denoted with $V_i = \{v_{i_1}, \ldots, v_{i_k}\}$, defining a partition of **V**.


### 1.1.3 Further Reading

For a broader coverage of the properties of directed and mixed graphs, we refer the reader to the monograph by Bang-Jensen and Gutin (2009), which at the time of this writing is the most complete reference on the subject. For undirected graphs, we refer to the classic book of Diestel (2005).


## 1.2 The R Environment for Statistical Computing

R (R Development Core Team, 2012) is a programming language and an environment targeted at statistical computing, released as an open-source software under the GNU General Public License (GPL). The main Web site of the R Project is http://www.r-project.org.

R supports all common operating systems (Windows, MacOS X and Linux) in addition to several Unix variants and has been constantly updated and improved over the years to become a standard choice for data analysis and the development of new statistical techniques.


### 1.2.1 Base Distribution and Contributed Packages

The R environment consists of a *base distribution*, maintained and developed by the R Core Team, and a constantly growing set of *contributed packages*. Both are distributed through a network of servers called "The Comprehensive R Archive Network" (CRAN), which provides up-to-date mirrors of the main site, located at http://cran.r-project.org.

The base distribution provides a set of standard packages implementing the basic functionality of R, including the following:

- Probability, density, distribution, and quantile functions for commonly used probability distributions

- Functions to produce nicely formatted plots such as boxplots, histograms, and scatterplots
- Statistical models such as linear and generalized linear models as well as functions for statistical hypothesis testing
- Several reference data sets from literature
- Utilities to import and export data in various formats (e.g., space- and tab-separated text; comma-separated values (CSV); files saves from other statistical software such as STATA, SPSS, and Octave; and many more).

Contributed packages are implemented by independent developers and then submitted to CRAN, which provides a unified distribution network and basic quality checking. In recent years it has become increasingly common to provide reference implementations of new methodologies as R packages. This trend has improved the reproducibility of scientific results presented in literature and, at the same time, has increased dramatically the number of fields in which R is a valuable data analysis tool.

## 1.2.2 A Quick Introduction to R

We will now illustrate some basic R commands for importing, exploring, summarizing, and plotting data. For this purpose, we will use the `lizards` data set included in the **bnlearn** package because of its simple structure. This data set was originally published in Schoener (1968) and has been used by Fienberg (1980) and more recently by Edwards (2000) as an example in the respective books.

First of all, we need to install the **bnlearn** package from one of the mirrors of the CRAN network. After launching R, we can type the following command after the ">" prompt:

```
> install.packages("bnlearn")
```

An up-to-date list of mirrors to choose from will be displayed as either a pop-up window or a text prompt. Once **bnlearn** has been installed, it can be loaded with

```
> library(bnlearn)
```

Clearly, `install.package` needs to be called only once for any given package, while loading the package with `library` is required at every new R session even when the workspace of the last session has been restored at start-up.

The `lizards` data set can then be loaded from **bnlearn** with

```
> data(lizards)
```

since the package is now loaded in the R session. If the data were stored in a text file, we could have imported them into R using the `read.table` function as follows:

```
> lizards = read.table("lizards.txt", header = TRUE)
```

Setting the header argument to TRUE tells read.table that the first line of the file lizards.txt contains the variable names. Each observation must be written in a single line, and the values assumed by the variables for that observation correspond to the fields (separated by spaces or tabulations) present in that line.

In both cases, the data is stored in a *data frame* called lizards, whose structure can be examined with the str function.

```
> str(lizards)
'data.frame': 409 obs. of  3 variables:
 $ Species : Factor w/ 2 levels "Sagrei","Distichus":
 1 ...
 $ Diameter: Factor w/ 2 levels "narrow","wide": 1 ...
 $ Height  : Factor w/ 2 levels "high","low": 2 2 ...
```

Like most programming languages, R defines a large set of *classes* of objects, which represent and provide an interface to different types of variables. Some of these classes correspond to various kinds of variables used in statistical modeling:

- *Logical*: indicator variables, e.g., either TRUE or FALSE
- *Integer*: natural numbers, e.g., $1, 2, \ldots, n \in \mathbb{N}$
- *Numeric*: real numbers, such as $1.2, \pi, \sqrt{2}$
- *Character*: character strings, such as "a", "b", "c"
- *Factor*: categorical variables, defined over a finite set of *levels* identified by character strings
- *Ordered*: ordered categorical variables, similar to factors but with an explicit ordering of the levels, e.g., "LOW" < "AVERAGE" < "HIGH".

Other classes correspond to more complex data types, such as multidimensional or heterogeneous data:

- *List*: a collection of arbitrary objects, often belonging to different classes
- *Vector*: a mathematical vector of elements belonging to the same class (i.e., all integers, all factors with the same levels, etc.) with an arbitrary number of dimensions
- *Matrix*: a matrix (i.e., a 2-dimensional vector) of elements belonging to the same class
- *Data frame*: a list of objects with the same length but possibly different classes. It is usually displayed and manipulated in the same way as a matrix.

As we can see from the output of str, read.table saves the data read from lizards.txt in a data frame to allow each variable to be stored as an object of the appropriate class. The labels of the possible values of each variable, which are character strings in lizards.txt, are automatically used as levels and the variables converted to factors.

We can further investigate the characteristics of the lizards data frame with the summary and dim functions.

```
> summary(lizards)
      Species         Diameter      Height
  Sagrei    :164    narrow:252    high:264
  Distichus:245    wide  :157    low :145
> dim(lizards)
[1] 409    3
```

From the output of str, summary, and dim, we can see that the data frame contains 409 observations and 3 variables named Species, Diameter, and Height. Each observation refers to a single lizard and describes its species (either *sagrei* or *distichus*) and the height and width of the branch it was perched on when sighted. All the variables are categorical and therefore are stored as *factors*; the values they can assume can be listed with the levels function.

```
> levels(lizards[, "Species"])
[1] "Sagrei"    "Distichus"
> levels(lizards[, "Height"])
[1] "high" "low"
> levels(lizards[, "Diameter"])
[1] "narrow" "wide"
```

An alternative, useful way of displaying these data is a contingency table, which can be built using the table function.

```
> table(lizards[, c(3, 2, 1)])
, , Species = Sagrei

        Diameter
Height narrow wide
   high     86   35
   low      32   11

, , Species = Distichus

        Diameter
Height narrow wide
   high     73   70
   low      61   41
```

The order in which the two-dimensional contingency tables are listed depends on the order of the variables in the data frame; in this case it is useful to have them split by specie first, so the columns of lizards were rearranged appropriately.

Exploratory data analysis often includes some form of graphical data visualization, especially when dealing with low-dimensional data sets such as the one we

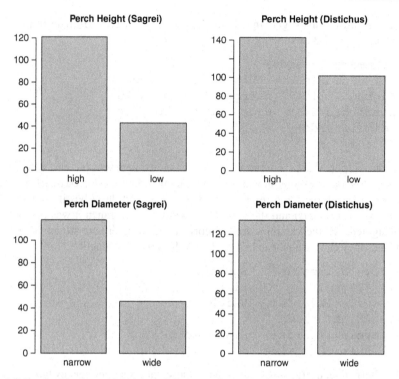

**Fig. 1.3** Barplots for the perch height and diameter of *sagrei* and *distichus* lizards

are considering. A simple way to plot the frequencies associated with `Height` and `Diameter` for each species is to use a *barplot*.

```
> Sagrei.lizards =
+       lizards[lizards$Species == "Sagrei", ]
> Distichus.lizards =
+       lizards[lizards$Species == "Distichus", ]
> par(mfrow = c(2, 2))
> plot(Sagrei.lizards[, "Height"],
+       main = "Perch Height (Sagrei)")
> plot(Distichus.lizards[, "Height"],
+       main = "Perch Height (Distichus)")
> plot(Sagrei.lizards[, "Diameter"],
+       main = "Perch Diameter (Sagrei)")
> plot(Distichus.lizards[, "Diameter"],
+       main = "Perch Diameter (Distichus)")
```

Figure 1.3 shows the plot generated be the commands above. The first two commands extract from the data set the subsets of observations corresponding to each species. `par` is then used to split the plot area into four quadrants, arranged in a layout with 2 rows and 2 columns. Each quadrant holds one of the barplots, which are generated by the `plot` function. `plot` is a generic function for data visualization that chooses a suitable plot depending on the data, in this case, barplots for factors. The `main` argument specifies the title of each plot, while `xlab` and `ylab` specify the labels to the horizontal and vertical axes, respectively.

Exploring numeric data requires many of the R functions illustrated above for categorical data. According to the description of the `lizards` data provided in Schoener (1968), a branch is classified as `narrow` if its diameter is lesser or equal than 4 inches and `wide` otherwise. For the sake of the example, we can generate some random values according to these specifications and associate them with the `Species`.

```
> diam = numeric(length = nrow(lizards))
> narrow = (lizards$Diameter == "narrow")
> wide = (lizards$Diameter == "wide")
> diam[narrow] = runif(n = 252, min = 2, max = 4)
> diam[wide] = runif(n = 157, min = 4, max = 6)
> new.data = data.frame(
+                 Species = lizards[, "Species"],
+                 Sim.Diameter = diam)
```

First, we create a new vector called `diam` with one entry for each observation (the number of rows (`nrow`) of `lizards`). Then we create two logical vectors with indicator variables identifying which branches are `narrow` and which are `wide` and use it to correctly assign the generated random diameters. The `runif` function generates independent random values from a uniform distribution in the range $(2, 4)$ for `narrow` branches and in $(4, 6)$ for `wide` branches. The (now populated) vector `diam` is then stored as `Sim.Diameter` in a new data frame called `new.data` along with the `Species` variable from the original data.

The behavior of `Sim.Diameter` can again be examined using `summary`, which in this case reports the mean and the quantiles of the new variable.

```
> summary(new.data[, "Sim.Diameter"])
   Min. 1st Qu.  Median    Mean 3rd Qu.    Max.
  2.000   2.855   3.641   3.759   4.640   5.982
```

It is also interesting to investigate how the diameter differs between the two `Species`, both in location (with `summary` again) and variability (with `var`).

```
> is.sagrei = (new.data[, "Species"] == "Sagrei")
> summary(new.data[is.sagrei, "Sim.Diameter"])
   Min. 1st Qu.  Median    Mean 3rd Qu.    Max.
  2.000   2.777   3.467   3.566   4.190   5.965

> summary(new.data[!is.sagrei, "Sim.Diameter"])
   Min. 1st Qu.  Median    Mean 3rd Qu.    Max.
  2.035   2.903   3.807   3.888   4.838   5.982
> var(new.data[is.sagrei, "Sim.Diameter"])
[1] 1.038415
```

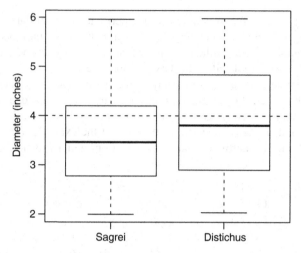

**Fig. 1.4** Boxplots for the simulated branch diameters for *sagrei* and *distichus* lizards

```
> var(new.data[!is.sagrei, "Sim.Diameter"])
[1] 1.333333
```

The same comparison can be performed graphically by plotting the *boxplots* corresponding to the diameters of the branches for each specie next to each other; the resulting plot is displayed in Fig. 1.4.

```
> boxplot(Sim.Diameter ~ Species, data = new.data,
+       ylab = "Diameter (inches)")
> abline(h = 4, lty = "dashed")
```

The call to `abline` adds a reference line separating `narrow` branches from `wide` ones, which helps relating the behavior of the new data to the original ones. The `lty` parameter specifies the line type, which in this case is dashed.

### 1.2.3 Further Reading

Providing a complete introduction to the R language in these few pages is clearly impossible, and it is outside the scope of this book. R and CRAN provide a comprehensive documentation in the form of several manuals available from CRAN and help pages distributed with the base and contributed packages describing the characteristics of each function and data set. Help pages can be accessed from within R using the `help`, `help.search` functions and the ? operator (e.g., `?runif`).

We refer the reader to the excellent "Modern Applied Statistics with S" by Venables and Ripley (2002) and to "Data Manipulation with R" by Spector (2009) for an in-depth coverage of R's capabilities. Readers interested in understanding the finer points of R programming, such as performance tuning and integrating compiled code, should also check "S Programming" by Venables and Ripley (2000).

## Exercises

**1.1.** Consider a directed acyclic graph with $n$ nodes.

(a) Show that at least one node must not have any incoming arc, i.e., the graph must contain at least one root node.
(b) Show that such a graph can have at most $\frac{1}{2}n(n-1)$ arcs.
(c) Show that a path can span at most $n-1$ arcs.
(d) Describe an algorithm to determine the topological ordering of the graph.

**1.2.** Consider the graphs shown in Fig. 1.1.

(a) Obtain the skeleton of the partially directed and directed graphs.
(b) Enumerate the acyclic graphs that can be obtained by orienting the undirected arcs of the partially directed graph.
(c) List the arcs that can be reversed (i.e., turned in the opposite direction), one at a time, without introducing cycles in the directed graph.

**1.3.** The (famous) `iris` data set reports the measurements in centimeters of the sepal length and width and the petal length and width for 50 flowers from each of 3 species of iris ("setosa," "versicolor," and "virginica").

(a) Load the `iris` data set (it is included in the **datasets** package, which is part of the base R distribution and does not need to be loaded explicitly) and read its manual page.
(b) Investigate the structure of the data set.
(c) Compare the sepal length among the three species by plotting histograms side by side.
(d) Repeat the previous point using boxplots.

**1.4.** Consider again the `iris` data set from Exercise 1.3.

(a) Write the data frame holding `iris` data frame into a space-separated text file named "iris.txt," and read it back into a second data frame called `iris2`.
(b) Check that `iris` and `iris2` are identical.
(c) Repeat the previous two steps with a file compressed with `bzip2` named "iris.txt.bz2."
(d) Save `iris` directly (e.g., without converting it to a text table) into a file called "iris.rda," and read it back.
(e) List all R objects in the global environment and remove all of them apart from `iris`.
(f) Exit the R saving the contents of the current session.

**1.5.** Consider the `gaussian.test` data set included in **bnlearn**.

(a) Print the column names.
(b) Print the range and the quartiles of each variable.
(c) Print all the observations for which A falls in the interval $[3,4]$ and B in $(-\infty, -5] \cup [10, \infty)$.
(d) Sample 50 rows without replacement.
(e) Draw a bootstrap sample (e.g., sample $5,000$ observations with replacement) and compute the mean of each variable.
(f) Standardize each variable.

**1.6.** Generate a data frame with 100 observations for the following variables:

(a) A categorical variable with two levels, low and high. The first 50 observations should be set to low, the others to high.
(b) A categorical variable with two levels, good and bad, nested within the first variable, i.e., the first 25 observations should be set to good, the second 25 to bad, and so on.
(c) A continuous, numerical variable following a Gaussian distribution with mean 2 and variance 4 when the first variable is equal to low and with mean 4 and variance 1 if the first variable is equal to high.

In addition, compute the standard deviation of the last variable for each configuration of the first two variables.

# Chapter 2
# Bayesian Networks in the Absence of Temporal Information

**Abstract** Data recorded across multiple variables of interest for a given phenomenon often do not contain any explicit temporal information. In the absence of such information, the data essentially represent a static snapshot of the underlying phenomenon at a particular moment in time. For this reason, they are sometimes referred to as *static data*.

Static Bayesian networks, commonly known simply as Bayesian networks, provide an intuitive and comprehensive framework to model the dependencies between the variables in static data. In this chapter, we will introduce the essential definitions and properties of static Bayesian networks. Subsequently, we will discuss existing Bayesian network learning algorithms and illustrate their applications with real-world examples and different R packages.

## 2.1 Bayesian Networks: Essential Definitions and Properties

Bayesian networks are a class of *graphical models* that allow a concise representation of the probabilistic dependencies between a given set of random variables $\mathbf{X} = \{X_1, X_2, \ldots, X_p\}$ as a *directed acyclic graph* (DAG) $G = (\mathbf{V}, A)$. Each node $v_i \in \mathbf{V}$ corresponds to a random variable $X_i$.

### 2.1.1 Graph Structure and Probability Factorization

The correspondence between the graphical separation ($\perp\!\!\!\perp_G$) induced by the absence of a particular arc and probabilistic independence ($\perp\!\!\!\perp_P$) provides a convenient way to represent the dependencies between the variables. Such a correspondence is formally known as an *independency map* (Pearl, 1988) and is defined as follows.

**Definition 2.1 (Maps).** A graph $G$ is an *independency map* (I-map) of the probabilistic dependence structure $P$ of $\mathbf{X}$ if there is a one-to-one correspondence between

R. Nagarajan et al., *Bayesian Networks in R: with Applications in Systems Biology*,
Use R! 48, DOI 10.1007/978-1-4614-6446-4_2,
© Springer Science+Business Media New York 2013

the random variables in **X** and the nodes **V** of $G$, such that for all disjoint subsets **A**, **B**, **C** of

$$A \perp\!\!\!\perp_P B \mid C \Longleftarrow A \perp\!\!\!\perp_G B \mid C. \tag{2.1}$$

Similarly, $G$ is a *dependency map* (D-map) of $P$ if **X** we have

$$A \perp\!\!\!\perp_P B \mid C \Longrightarrow A \perp\!\!\!\perp_G B \mid C. \tag{2.2}$$

$G$ is said to be a *perfect map* of $P$ if it is both a D-map and an I-map,

$$A \perp\!\!\!\perp_P B \mid C \Longleftrightarrow A \perp\!\!\!\perp_G B \mid C, \tag{2.3}$$

and in this case, $P$ is said to be *isomorphic* or *faithful* to $G$.

The correspondence between the structure of the DAG $G$ and the conditional independence relationships it represents is elucidated by the *directed separation* criterion (Pearl, 1988), or *d-separation*, as discussed below.

**Definition 2.2 (D-separation).** If **A**, **B**, and **C** are three disjoint subsets of nodes in a DAG $G$, then **C** is said to *d-separate* **A** from **B**, denoted $A \perp\!\!\!\perp_G B \mid C$, if along every sequence of arcs[1] between a node in **A** and a node in **B** there is a node $v$ satisfying one of the following two conditions:

1. $v$ has converging arcs (i.e., there are two arcs pointing to $v$ from the adjacent nodes in the path) and none of $v$ or its descendants (i.e., the nodes that can be reached from $v$) are in **C**.
2. $v$ is in **C** and does not have converging arcs.

The *Markov property* of Bayesian networks, which follows directly from d-separation, enables the representation of the joint probability distribution of the random variables in **X** (the *global distribution*) as a product of conditional probability distributions (the *local distributions* associated with each variable $X_i$). This is a direct application of the *chain rule* (Korb and Nicholson, 2010). In the case of discrete random variables, the factorization of the joint probability distribution $P_\mathbf{X}$ is given by

$$P_\mathbf{X}(\mathbf{X}) = \prod_{i=1}^{p} P_{X_i}(X_i \mid \Pi_{X_i}), \tag{2.4}$$

where $\Pi_{X_i}$ is the set of the parents of $X_i$; in the case of continuous random variables, the factorization of the joint density function $f_\mathbf{X}$ is given by

$$f_\mathbf{X}(\mathbf{X}) = \prod_{i=1}^{p} f_{X_i}(X_i \mid \Pi_{X_i}). \tag{2.5}$$

Similar results hold for mixed probability distributions (i.e., probability distributions including both discrete and continuous random variables).

---

[1] They are often referred to as *paths*, using the more general definition that disregards arc directions.

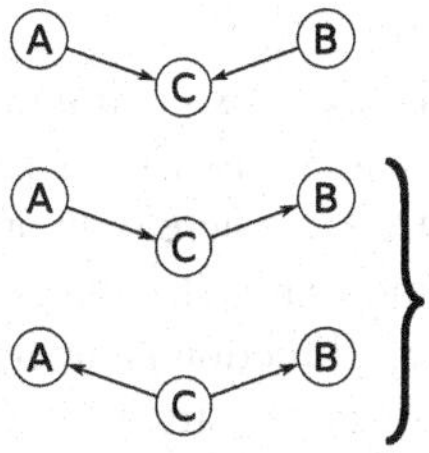

$$A \not\perp_G B \mid C$$
$$P(A, B, C) = P(C \mid A, B)\, P(A)\, P(B)$$

$$A \perp_G B \mid C \Rightarrow A \perp_P B \mid C$$
$$P(A, B, C) =$$
$$= P(B \mid C)\, P(C \mid A)\, P(A)$$
$$= P(A \mid C)\, P(B \mid C)\, P(C)$$

**Fig. 2.1** Graphical separation, conditional independence, and probability decomposition for the three *fundamental connections* (from *top* to *bottom*): *converging connection, serial connection,* and *diverging connection*

## 2.1.2 Fundamental Connections

Consider the *fundamental connections* (Jensen, 2001) shown in Fig. 2.1, the three possible configurations of three nodes and two arcs. In the *convergent connection* or *v-structure*, node $C$ has incoming arcs from $A$ and $B$, thus violating both conditions in Definition 2.2. Therefore, we conclude that $C$ does not d-separate $A$ and $B$. This in turn implies that $A$ and $B$ are not independent given $C$, and since $\Pi_A = \{\varnothing\}$, $\Pi_B = \{\varnothing\}$, and $\Pi_C = \{A, B\}$, we have

$$P(A, B, C) = P(C \mid A, B)\, P(A)\, P(B) \tag{2.6}$$

from the Markov property introduced in Eq. 2.4. From the above expression, it is evident that $C$ depends on the joint distributions of $A$ and $B$. Therefore, $A$ and $B$ are not conditionally independent given $C$. On the other hand, $A$ and $B$ are independent given $C$ in the *serial* and *diverging* connections since the conditions in Definition 2.2 are satisfied in these cases. For the serial connection, we have $\Pi_A = \{\varnothing\}$, $\Pi_B = \{C\}$, and $\Pi_C = \{A\}$; therefore,

$$P(A, B, C) = P(B \mid C)\, P(C \mid A)\, P(A). \tag{2.7}$$

For the diverging connection, we have $\Pi_A = \{C\}$, $\Pi_B = \{C\}$, and $\Pi_C = \{\varnothing\}$; therefore,

$$P(A, B, C) = P(A \mid C)\, P(B \mid C)\, P(C). \tag{2.8}$$

## 2.1.3 Equivalent Structures

From Fig. 2.1, it should also be noted that the *serial* and *diverging* connections result in equivalent factorizations; each can be obtained from the other with repeated applications of Bayes' theorem. Such probabilistically equivalent structures are known as *Markov equivalent* structures. Since equivalence is symmetric, reflexive, and transitive, each set of equivalent structures forms an *equivalence class*.

Generalizing this simple example, it can be shown that the only arcs whose direction is needed to identify an equivalence class are those belonging to at least one v-structure (Chickering, 1995).[2] Equivalence classes are usually represented by *completed partially directed acyclic graphs* (CPDAGs), where only arcs belonging to v-structures and those that would introduce additional v-structures or cycles are directed. Such arcs are called *compelled*, since their direction is determined by the equivalence class even though they are not part of any v-structure. Changing the direction of any other, non-compelled arc results in another network in the same equivalence class as long as it does not introduce any new v-structure or in any cycle.

### 2.1.4 Markov Blankets

Another fundamental quantity that is closely related to Definitions 2.1 and 2.2 is the *Markov blanket* (Pearl, 1988). It essentially represents the set of nodes that completely d-separates a given node from the rest of the graph.

**Definition 2.3 (Markov blanket).** The *Markov blanket* of a node $A \in \mathbf{V}$ is the minimal subset $\mathbf{S}$ of $\mathbf{V}$ such that

$$A \perp\!\!\!\perp_P \mathbf{V} - \mathbf{S} - A \,|\, \mathbf{S}. \tag{2.9}$$

In any Bayesian network, the Markov blanket of a node $A$ is the set of the parents of $A$, the children of $A$, and all the other nodes sharing a child with $A$.

Markov blankets facilitate the comparison of Bayesian networks with graphical models based on undirected graphs, which are known as *Markov networks* or *Markov random fields* (Whittaker, 1990; Edwards, 2000). On a related note, a DAG can be transformed in the undirected graph of the corresponding Markov networks by following the steps below.

1. Connect the nonadjacent nodes in each v-structure with an undirected arc. This is equivalent to adding an undirected arc between any node in the Markov blanket and the node the Markov blanket is centered on.
2. Ignore the direction of the other arcs. This effectively replaces the arcs with edges.

The above transformation is called *moralization* since it "marries" nonadjacent parents sharing a common child. The resulting graph is called a *moral graph* (Castillo et al., 1997).

---

[2] Note that the two parents in a v-structure ($A$ and $B$ in Fig. 2.1) cannot be connected by an arc, while this is not necessarily the case in a convergent connection.

---

**Algorithm 2.1** Inductive Causation Algorithm

---

1. For each pair of variables $A$ and $B$ in $\mathbf{V}$ search for set $\mathbf{S}_{AB} \subset V$ (including $S = \varnothing$) such that $A$ and $B$ are independent given $\mathbf{S}_{AB}$ and $A, B \notin \mathbf{S}_{AB}$. If there is no such a set, place an undirected arc between $A$ and $B$.
2. For each pair of non-adjacent variables $A$ and $B$ with a common neighbor $C$, check whether $C \in \mathbf{S}_{AB}$. If this is not true, set the direction of the arcs $A - C$ and $C - B$ to $A \rightarrow C$ and $C \leftarrow B$.
3. Set the direction of arcs which are still undirected by applying recursively the following two rules:

   a. if $A$ is adjacent to $B$ and there is a strictly directed path from $A$ to $B$ (a path leading from $A$ to $B$ containing no undirected arcs) then set the direction of $A - B$ to $A \rightarrow B$;
   b. if $A$ and $B$ are not adjacent but $A \rightarrow C$ and $C - B$, then change the latter to $C \rightarrow B$.

4. Return the resulting (completed partially) directed acyclic graph.

---

## 2.2 Static Bayesian Networks Modeling

The task of fitting a Bayesian network is usually called *learning*, a term borrowed from expert systems theory and artificial intelligence (Koller and Friedman, 2009). It is performed in two different steps, which correspond to model selection and parameter estimation techniques in classic statistical models.

The first step is called *structure learning* and consists in identifying the graph structure of the Bayesian network. Ideally, it should be the minimal I-map of the dependence structure of the data or, failing that, it should at least result in a distribution as close as possible to the correct one in the probability space. Several algorithms have been proposed in the literature for structure learning. Despite the variety of theoretical backgrounds and terminology, they fall under three broad categories: *constraint-based*, *score-based*, and *hybrid* algorithms. As an alternative, the network structure can be built manually from the domain knowledge of a human expert and prior information available on the data.

The second step is called *parameter learning*. As the name suggests, it implements the estimation of the parameters of the global distribution. This task can be performed efficiently by estimating the parameters of the local distributions implied by the structure obtained in the previous step.

### 2.2.1 Constraint-Based Structure Learning Algorithms

Constraint-based structure learning algorithms are based on the seminal work of Pearl on maps and its application to causal graphical models. His *inductive causation* (IC) algorithm (Verma and Pearl, 1991) provides a framework for learning the structure of Bayesian networks using conditional independence tests.

The details of the IC algorithm are described in Algorithm 2.1. The first step identifies which pairs of variables are connected by an arc, regardless of its direction. These variables cannot be independent given any other subset of variables, because they cannot be d-separated. This step can also be seen as a backward selection procedure starting from the saturated model with a complete graph and pruning it based on statistical tests for conditional independence.

The second step deals with the identification of the v-structures among all the pairs of nonadjacent nodes $A$ and $B$ with a common neighbor $C$. By definition, v-structures are the only fundamental connection in which the two nonadjacent nodes are not independent conditional on the third one. Therefore, if there is a subset of nodes that contains $C$ and d-separates $A$ and $B$, the three nodes are part of a v-structure centered on $C$. This condition can be verified by performing a conditional independence test for $A$ and $B$ against every possible subset of their common neighbors that includes $C$. At the end of the second step, both the skeleton and the v-structures of the network are known, so the equivalence class the Bayesian network belongs to is uniquely identified.

The third and last step of the IC algorithm identifies compelled arcs and orients them recursively to obtain the completed partially DAG (CPDAG) describing the equivalence class identified by the previous steps.

A major problem of the IC algorithm is that the first two steps cannot be applied in the form described in Algorithm 2.1 to any real-world problem due to the exponential number of possible conditional independence relationship. This has led to the development of improved algorithms such as the following:

- *PC*: the first practical application of the IC algorithm (Spirtes et al., 2001), a backward selection procedure from the saturated graph
- *Grow-Shrink* (GS): based on the *Grow-Shrink Markov blanket* algorithm (Margaritis, 2003), a simple forward selection Markov blanket detection approach
- *Incremental Association* (IAMB): based on the *Incremental Association Markov blanket* algorithm (Tsamardinos et al., 2003), a two-phase selection scheme based on a forward selection followed by a backward one
- *Fast Incremental Association* (Fast-IAMB): a variant of IAMB which uses speculative stepwise forward selection to reduce the number of conditional independence tests (Yaramakala and Margaritis, 2005)
- *Interleaved Incremental Association* (Inter-IAMB): another variant of IAMB which uses forward stepwise selection (Tsamardinos et al., 2003) to avoid false positives in the Markov blanket detection phase

All these algorithms, with the exception of PC, first learn the Markov blanket of each node in the network. This preliminary step greatly simplifies the identification of neighbors of each node, as the search can be limited to its Markov blanket. As a result, the number of conditional independence tests performed by the learning algorithm and its overall computational complexity are significantly reduced.

---
**Algorithm 2.2** Hill-Climbing Algorithm
---

1. Choose a network structure $G$ over $\mathbf{V}$, usually (but not necessarily) empty.
2. Compute the score of $G$, denoted as $Score_G = \text{Score}(G)$.
3. Set *maxscore* $= Score_G$.
4. Repeat the following steps as long as *maxscore* increases:

   a. for every possible arc addition, deletion or reversal not resulting in a cyclic network:
      i. compute the score of the modified network $G^*$, $Score_{G^*} = \text{Score}(G^*)$:
      ii. if $Score_{G^*} > Score_G$, set $G = G^*$ and $Score_G = Score_{G^*}$.
   b. update *maxscore* with the new value of $Score_G$.

5. Return the directed acyclic graph $G$.

---

Further improvements are possible by leveraging the symmetry of Markov blankets implied in Definition 2.3 and shown in Sect. 2.3.

## 2.2.2 Score-Based Structure Learning Algorithms

Score-based structure learning algorithms (also known a *search-and-score algorithms*) represent the application of general heuristic optimization techniques to the problem of learning the structure of a Bayesian network. Each candidate network is assigned a *network score* reflecting its goodness of fit, which the algorithm then attempts to maximize. Some examples from this class of algorithms are the following:

- *Greedy search* algorithms such as *hill-climbing* with *random restarts* or *tabu search* (Bouckaert, 1995). These algorithms explore the search space starting from a network structure (usually the empty graph) and adding, deleting, or reversing one arc at a time until the score can no longer be improved (see Algorithm 2.2).
- *Genetic* algorithms, which mimic natural evolution through the iterative selection of the "fittest" models and the hybridization of their characteristics (Larrañaga et al., 1997). In this case the search space is explored through the *crossover* (which combines the structure of two networks) and *mutation* (which introduces random alterations) stochastic operators.
- *Simulated annealing* (Bouckaert, 1995). This algorithm performs a stochastic local search by accepting changes that increase the network score and, at the same time, allowing changes that decrease it with a probability inversely proportional to the score decrease.

A comprehensive review of these heuristics, as well as related approaches from the field of artificial intelligence, is provided in Russell and Norvig (2009).

---

**Algorithm 2.3** Sparse Candidate Algorithm

---

1. Choose a network structure $G$ over $\mathbf{V}$, usually (but not necessarily) empty.
2. Repeat the following steps until convergence:

   a. **restrict:** select a set $\mathbf{C}_i$ of candidate parents for each node $X_i \in \mathbf{V}$, which must include the parents of $X_i$ in $G$;
   b. **maximize:** find the network structure $G^*$ that maximizes $\mathrm{Score}(G^*)$ among the networks in which the parents of each node $X_i$ are included in the corresponding set $\mathbf{C}_i$;
   c. set $G = G^*$.

3. Return the directed acyclic graph $G$.

---

### 2.2.3 Hybrid Structure Learning Algorithms

Hybrid structure learning algorithms combine constraint-based and score-based algorithms to offset their weaknesses and produce reliable network structures in a wide variety of situations. The two best-known members of this family are the *Sparse Candidate* algorithm (SC) by Friedman et al. (1999b) and the *Max-Min Hill-Climbing* (MMHC) algorithm by Tsamardinos et al. (2006). The former is illustrated in Algorithm 2.3.

Both these algorithms are based on two steps called *restrict* and *maximize*. In the first one, the candidate set for the parents of each node $X_i$ is reduced from the whole node set $\mathbf{V}$ to a smaller set $C_i \subset \mathbf{V}$ of nodes whose behavior has been shown to be related in some way to that of $X_i$. This in turn results in a smaller and more regular search space. The second step seeks the network that maximizes a given score function, subject to the constraints imposed by the $\mathbf{C}_i$ sets.

In the Sparse Candidate algorithm these two steps are applied iteratively until there is no change in the network or no network improves the network score; the choice of the heuristics used to perform them is left to the implementation. On the other hand, in the MMHC algorithm, *restrict* and *maximize* are performed only once; the *Max-Min Parents and Children* (MMPC) heuristic is used to learn the candidate sets $\mathbf{C}_i$ and a hill-climbing greedy search to find the optimal network.

### 2.2.4 Choosing Distributions, Conditional Independence Tests, and Network Scores

In principle, there are many possible choices for both the global and the local distribution functions, depending on the nature of the data and the aims of the analysis. However, literature has focused mostly on two cases:

- *Multinomial variables*: used for discrete/categorical data sets and often referred to as the *discrete case*. Both the global and the local distributions are multinomial, and the latter are represented as *conditional probability tables* (CPTs). This is by far the most common assumption in literature, and the corresponding Bayesian networks are referred to as *discrete Bayesian networks*.
- *Multivariate normal variables*: this representation is used for continuous data sets and is therefore referred to as the *continuous case*. The global distribution is multivariate normal, whereas the local distributions are univariate normal random variables linked by linear constraints. Local distributions are in fact linear models in which the parents play the role of explanatory variables. These Bayesian networks are called *Gaussian Bayesian networks* (Geiger and Heckerman, 1994; Neapolitan, 2003).

Other distributional assumptions require ad hoc learning algorithms or present various limitations due to the difficulty of specifying the distribution functions in closed form. For example, models for mixed data, such as the one presented in Bøttcher and Dethlefsen (2003), impose constraints on the choice of the parents for the nodes.

On a related note, the choice of a particular set of global and local distributions also determines which conditional independence tests and which network scores can be used to learn the structure of the Bayesian network.

Conditional independence tests and network scores for discrete data are functions of the CPTs implied by the graphical structure of the network through the observed frequencies $\{n_{ijk}, i = 1, \ldots, R, j = 1, \ldots, C, k = 1, \ldots, L\}$ for the random variables $X$ and $Y$ and all the configurations of the conditioning variables $\mathbf{Z}$. Two common conditional independence tests are the following:

- *Mutual information* (Cover and Thomas, 2006), an information-theoretic distance measure defined as

$$\mathrm{MI}(X, Y \mid \mathbf{Z}) = \sum_{i=1}^{R} \sum_{j=1}^{C} \sum_{k=1}^{L} \frac{n_{ijk}}{n} \log \frac{n_{ijk} n_{++k}}{n_{i+k} n_{+jk}}. \tag{2.10}$$

It is proportional to the log-likelihood ratio test $G^2$ (they differ by a $2n$ factor, where $n$ is the sample size), and it is related to the deviance of the tested models.

- The classic *Pearson's $X^2$* test for contingency tables,

$$X^2(X, Y \mid \mathbf{Z}) = \sum_{i=1}^{R} \sum_{j=1}^{C} \sum_{k=1}^{L} \frac{(n_{ijk} - m_{ijk})^2}{m_{ijk}}, \quad \text{where} \quad m_{ijk} = \frac{n_{i+k} n_{+jk}}{n_{++k}}. \tag{2.11}$$

In both cases the null hypothesis of independence can be tested using either the asymptotic $\chi^2_{(R-1)(C-1)L}$ distribution or the Monte Carlo permutation approach described in Edwards (2000). Other possible choices are Fisher's exact test and the shrinkage estimator for the mutual information defined by Hausser and Strimmer (2009) and studied in Scutari and Brogini (2012).

Network scores commonly found in literature are the following:

- The *Bayesian Dirichlet equivalent* (BDe) score, the posterior density associated with a uniform prior over both the space of the network structures and of the parameters of each local distribution (Heckerman et al., 1995).
- The *Bayesian information criterion* (BIC), a penalized likelihood score defined as

$$\text{BIC} = \sum_{i=1}^{n} \log P_{X_i}(X_i \mid \Pi_{X_i}) - \frac{d}{2} \log n, \tag{2.12}$$

where $d$ is the number of parameters of the global distribution. It is numerically equivalent to the information-theoretic *minimum description length* (MDL) measure by Rissanen (2007), even though it has a completely different derivation. BIC converges asymptotically to the posterior density BDe.

These score functions are said to be *score equivalent*, since they assign the same score to networks belonging to the same equivalence class. They are also *decomposable* into the components associated with each node, which is a significant computational advantage when learning the structure of the network (the only parts of the score that need to be computed are those that differ between the networks being compared).

In the continuous case, conditional independence tests and network scores are functions of the partial correlation coefficients $\rho_{XY \mid \mathbf{Z}}$ of $X$ and $Y$ given $\mathbf{Z}$. Two common conditional independence tests are the following:

- The exact $t$ test for Pearson's correlation coefficient, defined as

$$\text{t}(X, Y \mid \mathbf{Z}) = \rho_{XY \mid \mathbf{Z}} \sqrt{\frac{n-2}{1 - \rho_{XY \mid \mathbf{Z}}^2}} \tag{2.13}$$

and distributed as a Student's $t$ with $n - |\mathbf{Z}| - 2$ degrees of freedom.
- *Fisher's Z* test, a transformation of the linear correlation coefficient with an asymptotic normal distribution and defined as

$$Z(X, Y \mid \mathbf{Z}) = \frac{\sqrt{n - |\mathbf{Z}| - 3}}{2} \log \frac{1 + \rho_{XY \mid \mathbf{Z}}}{1 - \rho_{XY \mid \mathbf{Z}}}. \tag{2.14}$$

Both tests can also be performed using Monte Carlo permutation approaches such as the ones described in Legendre (2000). Other possible choices are the mutual information test defined in Kullback (1968), which is proportional to the corresponding log-likelihood ratio test, or the shrinkage estimators developed by Shäfer and Strimmer (2005).

Commonly used network scores are again BIC, this time defined as

$$\text{BIC} = \sum_{i=1}^{n} \log f_{X_i}(X_i \mid \Pi_{X_i}) - \frac{d}{2} \log n \tag{2.15}$$

and the *Bayesian Gaussian equivalent* (BGe) score, the Wishart posterior density of the network associated with a uniform prior over both the space of the network structures and of the parameters of the local distributions (Geiger and Heckerman, 1994).

### 2.2.5 Parameter Learning

Once the structure of the network has been learned from the data, the task of estimating and updating the parameters of the global distribution is greatly simplified by the application of the Markov property.

Local distributions in practice involve only a small number of variables. Furthermore, their dimension usually does not scale with the size of **X** and is often assumed to be bounded by a constant when computing the computational complexity of algorithms. This in turn alleviates the *curse of dimensionality*, because each local distribution has a comparatively small number of parameters to estimate from the sample and because estimates are more accurate due to the better ratio between the size of parameter space and the sample size. There are two main approaches to the estimation of those parameters in literature: one based on *maximum likelihood estimation* and the other based on *Bayesian estimation*.

The number of parameters needed to uniquely identify the global distribution, which is the sum of the number of parameters of the local distributions, is also reduced because the conditional independence relationships encoded in the network structure fix large parts of the parameter space. For example, in Gaussian Bayesian networks, partial correlation coefficients involving (conditionally) independent variables are equal to zero by definition, and joint frequencies factorize into marginal ones in multinomial distributions.

However, parameter estimation is still problematic in many situations. For example, it is increasingly common to have sample sizes much smaller than the number of variables included in the model. This is typical of high-throughput biological data sets, such as microarrays, that have a few ten or hundred observations and thousands of genes. In this setting, which is called "small $n$, large $p$," estimates have a high variability unless particular care is taken in both structure and parameter learnings (Castelo and Roverato, 2006; Shäfer and Strimmer, 2005; Hastie et al., 2009).

### 2.2.6 Discretization

A simple way to learn Bayesian networks from mixed data is to convert all continuous variables to discrete ones and then to apply the techniques described in the previous sections. This approach, which is called *discretization* or *binning*, completely sidesteps the problem of defining a probabilistic model for the data. Discretization may also be applied to deal with continuous data when one or more variables present severe departures from normality (skewness, heavy tails, etc.).

The intervals the variables will be discretized into can be chosen in one of the following ways:

- Using prior knowledge on the data. The boundaries of the intervals are defined, for each variable, to correspond to significantly different real-world scenarios, such as the concentration of a particular pollutant (absent, dangerous, lethal) or age classes (child, adult, elderly).
- Using heuristics before learning the structure of the network. Some examples are Sturges, Freedman-Diaconis, or Scott rules (Venables and Ripley, 2002).
- Choosing the number of intervals and their boundaries to balance accuracy and information loss (Kohavi and Sahami, 1996), again one variable at a time and before the network structure has been learned. A similar approach considering pairs of variables is presented in Hartemink (2001).
- Performing learning and discretization iteratively until no improvement is made (Friedman and Goldszmidt, 1996).

These strategies represent different trade-offs between the accuracy of the discrete representation of the original data and the computational efficiency of the transformation.

## 2.3 Static Bayesian Networks Modeling with R

In this section, we demonstrate structure learning, parameter learning, and manipulation of a static Bayesian network in the R environment. Several of the packages introduced in Sect. 2.3.1 will be covered to provide an overview of the possibilities offered by the R environment. All code will be illustrated using a very simple data set and explained step by step to develop a throughout understanding of Bayesian network learning.

### 2.3.1 Popular R Packages for Bayesian Network Modeling

There are several packages on CRAN dealing with Bayesian networks. They can be divided in two categories: those that deal with structure learning and those that focus only on parameter learning and inference (Table 2.1).

Packages **bnlearn** (Scutari, 2010, 2012), **deal** (Bøttcher and Dethlefsen, 2003), **pcalg** (Kalisch et al., 2012), and **catnet** (Balov and Salzman, 2012) fall into the first category. **bnlearn** offers a wide variety of structure learning algorithms (spanning all the three classes covered in this chapter, with the tests and scores covered in Sect. 2.2.4), parameter learning approaches (maximum likelihood for discrete and continuous data, Bayesian estimation for discrete data), and inference tech-

**Table 2.1** Feature matrix for the R packages covered in Sect. 2.3.1

|                          | bnlearn | catnet | deal | pcalg | gRbase | gRain |
|--------------------------|---------|--------|------|-------|--------|-------|
| Discrete data            | Yes     | Yes    | Yes  | Yes   | Yes    | Yes   |
| Continuous data          | Yes     | No     | Yes  | Yes   | Yes    | No    |
| Mixed data               | No      | No     | Yes  | No    | No     | No    |
| Constraint-based learning| Yes     | No     | No   | Yes   | No     | No    |
| Score-based learning     | Yes     | Yes    | Yes  | No    | No     | No    |
| Hybrid learning          | Yes     | No     | No   | No    | No     | No    |
| Structure manipulation   | Yes     | Yes    | No   | No    | Yes    | No    |
| Parameter estimation     | Yes     | Yes    | Yes  | Yes   | No     | No    |
| Prediction               | Yes     | Yes    | No   | No    | No     | Yes   |
| Approximate inference    | Yes     | No     | No   | No    | No     | Yes   |

niques (cross-validation, bootstrap, conditional probability queries, and prediction). It is also the only package that keeps a clear separation between the structure of a network and the associated probability distribution, which are implemented as two different classes of R objects.

**deal** implements structure and parameter learning using a Bayesian approach and handles discrete, continuous, and mixed data (assuming a conditional Gaussian distribution). The network structure is learned with a hill-climbing greedy search such as the one described in Algorithm 2.2, with the posterior density of the network as a score function and random restarts to avoid local maxima.

**pcalg** provides a free software implementation of the PC algorithm and is specifically designed to estimate and measure causal effects. It handles both discrete and continuous data and can account for the effects of latent variables on the network. The latter is achieved through a modified PC algorithm known as *Fast Causal Inference* (FCI), first proposed by Spirtes et al. (2001).

**catnet** focuses on discrete, static Bayesian networks from a frequentist point of view. Structure learning is performed in two steps. First, the node ordering of the graph is learned from the data using simulated annealing; alternatively, a custom node ordering can be specified by the user. An exhaustive search is performed among the network structures with the given node ordering, and the exact maximum likelihood solution is returned. Parameter learning and prediction are also implemented. Furthermore, an extension of this approach for mixed data (assuming a Gaussian mixture distribution) has been recently made available from CRAN in package **mugnet** (Balov, 2011).

Packages **gRbase** (Højsgaard et al., 2010) and **gRain** (Højsgaard, 2010) fall into the second category. They focus on manipulating the parameters of the network, on prediction and on inference, under the assumption that all variables are discrete. Neither **gRbase** nor **gRain** implement any structure or parameter learning algorithm, so the Bayesian network must be completely specified by the user.

### 2.3.2 Creating and Manipulating Network Structures

Consider a data set consisting of the exam scores of 88 students across five different topics, namely, mechanics, vectors, algebra, analysis, and statistics. The scores are bounded in the interval $[0, 100]$. This data set was originally investigated by Mardia et al. (1979) and subsequently in classic books on graphical models such as Whittaker (1990) and Edwards (2000). A copy of the data is included in **bnlearn** under the name marks.

```
> library(bnlearn)
> data(marks)
> str(marks)
'data.frame':   88 obs. of  5 variables:
 $ MECH: num  77 63 75 55 63 53 51 59 62 64 ...
 $ VECT: num  82 78 73 72 63 61 67 70 60 72 ...
 $ ALG : num  67 80 71 63 65 72 65 68 58 60 ...
 $ ANL : num  67 70 66 70 70 64 65 62 62 62 ...
 $ STAT: num  81 81 81 68 63 73 68 56 70 45 ...
```

Upon loading the data, we can create an empty network with the nodes corresponding to the variables in marks using the empty.graph function.

```
> ug = empty.graph(names(marks))
```

We can then add the arcs present in the original network from Whittaker (1990) by assigning a two-column matrix containing the labels of their end-nodes. Undirected arcs are represented as their two possible orientations. For instance, the arc MECH − VECT is represented by the pair {MECH → VECT, VECT → MECH}.

```
> arcs(ug, ignore.cycles = TRUE) = matrix(
+    c("MECH", "VECT", "MECH", "ALG", "VECT", "MECH",
+       "VECT", "ALG", "ALG",  "MECH", "ALG", "VECT",
+       "ALG",  "ANL", "ALG",  "STAT", "ANL", "ALG",
+       "ANL",  "STAT", "STAT", "ALG", "STAT", "ANL"),
+    ncol = 2, byrow = TRUE,
+    dimnames = list(c(), c("from", "to")))
```

The resulting ug object belongs to the class bn, which is the S3 class used by the **bnlearn** package to manage network structures. It contains the following information (the element name is reported in parenthesis):

- learning: a list containing some information about the results of the structure learning algorithm and its tuning parameters, including the conditional independence tests and network scores used in the analysis. It has never changed after the object is created.
- nodes: a list containing one element per node. Each element contains the Markov blanket, the neighborhood, the parents, and the children of that particular node.

- arcs: the arcs present in the network, in the same two-column format used in the call to `arcs` above.

All these information can be accessed through ad hoc accessor functions, some of which will be illustrated in this section. Furthermore, a synthetic view of the network is provided by the `print` method for this class.

```
> ug

  Random/Generated Bayesian network

  model:
    [undirected graph]
  nodes:                                        5
  arcs:                                         6
    undirected arcs:                            6
    directed arcs:                              0
  average markov blanket size:                  2.40
  average neighbourhood size:                   2.40
  average branching factor:                     0.00

  generation algorithm:                         Empty
```

The structure of `ug` is shown in Fig. 2.2, along with one of the Bayesian networks that will be learned from `marks` in Sect. 2.3.4 and the corresponding equivalence class. As before, we can create a `bn` object for that Bayesian network with:

```
> dag = empty.graph(names(marks))
> arcs(dag) = matrix(
+     c("VECT", "MECH", "ALG", "MECH", "ALG", "VECT",
+       "ANL", "ALG", "STAT", "ALG", "STAT", "ANL"),
+     ncol = 2, byrow = TRUE,
+     dimnames = list(c(), c("from", "to")))

> dag

  Random/Generated Bayesian network

  model:
    [STAT] [ANL|STAT] [ALG|ANL:STAT] [VECT|ALG]
    [MECH|VECT:ALG]
  nodes:                                        5
  arcs:                                         6
    undirected arcs:                            0
    directed arcs:                              6
  average markov blanket size:                  2.40
  average neighbourhood size:                   2.40
```

Original Graphical Model (UG)        Bayesian Network (DAG)        Equivalence Class (CPDAG)

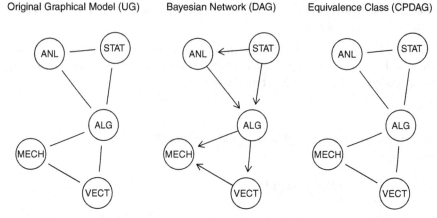

**Fig. 2.2** The undirected graphical model for the `marks` data set from Edwards (2000) and Whittaker (1990) (*left*), the Bayesian network learned from the same data (*center*), and the CPDAG of its equivalence class (*right*)

```
average branching factor:                    1.20

generation algorithm:                        Empty
```

Rather than typing the arcs of the network in a two-column format, we can also create ug starting from an adjacency matrix as follows:

```
> mat = matrix(c(0, 1, 1, 0, 0, 0, 0, 1, 0, 0, 0, 0,
+            0, 1, 1, 0, 0, 0, 0, 1, 0, 0, 0, 0),
+            nrow = 5,
+            dimnames = list(nodes(dag), nodes(dag)))

> mat
        MECH VECT ALG ANL STAT
MECH       0    0   0   0    0
VECT       1    0   0   0    0
ALG        1    1   0   0    0
ANL        0    0   1   0    0
STAT       0    0   1   1    0
> dag2 = empty.graph(nodes(dag))
> amat(dag2) = mat
> all.equal(dag, dag2)
[1] TRUE
```

On the other hand, sometimes we may just want to create a new bn object by modifying an existing one. The most straightforward way to accomplish this is by adding (set.arc), dropping (drop.arc), or reversing (rev.arc) arcs in the original network.

```
> dag3 = empty.graph(nodes(dag))
> dag3 = set.arc(dag3, "VECT", "MECH")
> dag3 = set.arc(dag3, "ALG", "MECH")
> dag3 = set.arc(dag3, "ALG", "VECT")
> dag3 = set.arc(dag3, "ANL", "ALG")
> dag3 = set.arc(dag3, "STAT", "ALG")
> dag3 = set.arc(dag3, "STAT", "ANL")
> all.equal(dag, dag3)
[1] TRUE
```

The approaches discussed above are guaranteed to result in directed or partially DAGs unless check.cycles is explicitly set to FALSE. A quick check reveals that the moral graph of dag and the graphical model from Whittaker (1990) express the same dependence relationships, as expected.

```
> all.equal(ug, moral(dag))
[1] TRUE
```

Upon creating a bn object, we are in a position to investigate those properties of the corresponding graph that have a probabilistic interpretation in a Bayesian network. For this purpose, the bn class provides a complete description of the network structure (which is uniquely specified by its arc set), and the use of the information stored for each node results in significant performance improvements for common operations.

For instance, when treating the network as a causal model we are often interested in the topological ordering of the nodes. The relative position of two nodes in the topological ordering is indicative of the direction of any possible causal relationship between them, because it implies the direction of any possible path linking the nodes (a more detailed explanation can be found in Sect. 2.4).

```
> node.ordering(dag)
[1] "STAT" "ANL"  "ALG"  "VECT" "MECH"
```

The neighborhood (nbr) and the Markov blanket (mb) of a node provide a synthetic description of the local dependence structure around that node. These can be obtained as follows:

```
> nbr(dag, "ANL")
[1] "ALG"  "STAT"
> mb(dag, "ANL")
[1] "ALG"  "STAT"
```

We can also use the commands above to show that both sets describe symmetric relationships, i.e., if ALG is in the Markov blanket of ANL, ANL is in the Markov blanket of ALG.

```
> "ANL" %in% mb(dag, "ALG")
[1] TRUE
> "ALG" %in% mb(dag, "ANL")
[1] TRUE
```

Furthermore, we can check that the Markov blanket of a given node (VECT in this example) is indeed composed by its children (chld), its parents (par), and its children's other parents (o.par), as stated in Definition 2.3.

```
> chld = children(dag, "VECT")
> par = parents(dag, "VECT")
> o.par = sapply(chld, parents, x = dag)
> unique(c(chld, par, o.par[o.par != "VECT"]))
[1] "MECH" "ALG"
> mb(dag, "VECT")
[1] "MECH" "ALG"
```

Several scoring criteria have been proposed in the context of structure learning. The example below demonstrates the log-likelihood score of a Bayesian network and how it changes in response to changes in the graph structure. More importantly, it demonstrates the invariance of the log-likelihood for networks in the same equivalence class as expected.

```
> score(dag, data = marks, type = "loglik-g")
[1] -1695.525
> dag.eq = reverse.arc(dag, "STAT", "ANL")
> score(dag.eq, data = marks, type = "loglik-g")
[1] -1695.525
```

As can be seen from Fig. 2.2, the arc STAT → ANL does not belong to any v-structure nor does ANL → STAT. This is easy to check using the vstructs function, which shows that in fact neither dag nor dag.eq contains any v-structure (denoted in the output a $X \rightarrow Z \leftarrow Y$).

```
> vstructs(dag)
      X Z Y
> vstructs(dag.eq)
      X Z Y
```

In all convergent connections present in dag and dag.eq, the two parent nodes are connected by an arc; therefore, they are not v-structures. Koller and Friedman (2009) call them *moral v-structures*, because the parents are "married" as in a moral graph.

```
> vstructs(dag, moral = TRUE)
       X         Z        Y
[1,]  "VECT"   "MECH"   "ALG"
[2,]  "ANL"    "ALG"    "STAT"
> vstructs(dag.eq, moral = TRUE)
       X         Z        Y
[1,]  "VECT"   "MECH"   "ALG"
[2,]  "ANL"    "ALG"    "STAT"
```

Score equivalence can be systematically checked by comparing the CPDAGs of the equivalence classes of dag and dag.eq, which can be derived using the cpdag function as shown below.

```
> all.equal(cpdag(dag), cpdag(dag.eq))
[1] TRUE
```

Similarly, we can derive the moral graphs of dag and dag.eq with moral and show them to be equal.

```
> all.equal(moral(dag), moral(dag.eq))
[1] TRUE
```

Of interest is to note that networks belonging to different equivalence classes may have the same moral graph but not vice versa. Consider, for instance, the networks shown in Fig. 2.3, obtained from dag by dropping, respectively, STAT → ANL and ALG → VECT.

```
> dag2 = drop.arc(dag, from = "STAT", to = "ANL")
> dag3 = drop.arc(dag, from = "ALG", to = "VECT")
```

dag2 and dag3 cannot belong to the same equivalence class because they contain different sets of v-structures, as shown below.

```
> vstructs(dag2)
      X      Z      Y
[1,] "ANL" "ALG" "STAT"
> vstructs(dag3)
      X      Z      Y
[1,] "VECT" "MECH" "ALG"
```

Equivalently, we can derive their CPDAGs and show them to be different as well.

```
> all.equal(cpdag(dag2), cpdag(dag3))
[1] "Different number of directed/undirected arcs"
```

However, dag2 and dag3 have identical moral graphs, and those moral graphs are identical to the moral graph of dag as well.

```
> all.equal(moral(dag2), moral(dag3))
[1] TRUE
> all.equal(moral(dag2), moral(dag))
[1] TRUE
> all.equal(moral(dag3), moral(dag))
[1] TRUE
```

All the examples covered above can be similarly implemented using the other packages described in Sect. 2.3.1. However, the lack of a clear separation between the handling of the network structure and the corresponding local distributions makes the analysis of the former more cumbersome. For example, both **deal** and **catnet** implement only a single object class (called network in **deal** and

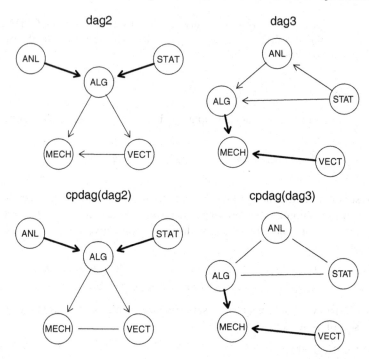

**Fig. 2.3** Two networks (dag2 and dag3) derived from dag, with different sets of v-structures and therefore belonging to different equivalence classes (cpdag(dag2) and cpdag(dag3)). Both these networks have the same moral graph as dag, shown in the *right panel* of Fig. 2.2. V-structures are highlighted with a *thicker line width*

catNetwork in **catnet**), describing a Bayesian network as a whole. This design choice makes network structures not as easy to modify as in **bnlearn**, because the parameters of the local distributions must be modified at the same time to preserve the coherence of the R object. Furthermore, in some cases the lack of accessor functions forces the user to work directly on the internals of the class, which increases the complexity of even simple tasks. On the other hand, class pcAlgo from **pcalg** stores the network structure in an object of class graphNEL, making it possible to use all the functions provided by the **graph** package (Gentleman et al., 2012).

Consider, for instance, the undirected graph and the DAG shown in Fig. 2.2. With the **deal** package we can again create an empty network, which in this case is an object of class network.

```
> library(deal)
> deal.net = network(marks)
> deal.net
## 5 ( 0 discrete+ 5 ) nodes;score=   ;relscore=
1    MECH continuous()
2    VECT continuous()
```

```
3    ALG continuous()
4    ANL continuous()
5    STAT continuous()
```

However, not only it is not possible to recreate the undirected graph from Whittaker (1990), but the only way to specify the DAG from Fig. 2.2 is through the same model string representation used in **bnlearn**.

```
> m = paste("[MECH] [VECT|MECH] [ALG|MECH:VECT]",
+                "[ANL|ALG] [STAT|ALG:ANL]", sep = "")
> deal.net = as.network(m, deal.net)
> deal.net
##   5 ( 0 discrete+ 5 ) nodes;score= NA ;relscore=
1    MECH continuous()
2    VECT continuous()    1
3    ALG continuous()     1    2
4    ANL continuous()     3
5    STAT continuous()    3    4
```

Package **catnet** on the other hand is able to import a network structure from any graphNEL object, making interoperation with **pcalg** easy, or through a list containing the parents of each node, as shown in the code below.

```
> library(catnet)
> cat.net = cnCatnetFromEdges(names(marks),
+                list(MECH = NULL, VECT = "MECH",
+                ALG = c("MECH", "VECT"), ANL = "ALG",
+                STAT = c("ALG", "ANL")))
> cat.net
A catNetwork object with  5  nodes,  2  parents,
  2  categories, Likelihood =  0 , Complexity =  13 .
```

For both packages, other quantities of interest have to be derived manually from the information stored in the respective classes. For example, the Markov blanket of VECT can be constructed in **catnet** using the functions cnEdges and cnParents as follows:

```
> chld = cnEdges(cat.net)$VECT
> par = cnParents(cat.net)$VECT
> o.par = sapply(chld,
+    function(node) { cnEdges(cat.net)[[node]] })
> unique(unlist(c(chld, par, o.par[o.par != "VECT"])))
[1] "MECH" "ALG"
```

Furthermore, `cnMatEdges` produces exactly the same output as the function `arcs` from **bnlearn**, making it easy to export models from the former to the latter.

```
> em = empty.graph(names(marks))
> arcs(em) = cnMatEdges(cat.net)
```

`modelstring` can be used in **deal** to the same effect.

```
> em = model2network(deal::modelstring(deal.net))
```

### 2.3.3 Plotting Network Structures

Visual examination of the graph structure may also provide useful insights into the properties of a Bayesian network, especially in the case of small- and medium-sized graphs. **bnlearn**, like many other packages dealing with graph structures, provides a set of plotting functions based on the interface provided by the **graph** and **Rgraphviz** (Gentry et al., 2012) packages. The options needed to reproduce the plots commonly found in literature are available through the `graphviz.plot` function, which takes a `bn` object as an argument and returns the corresponding `graph` object for further customizations. For example, the plots shown in Fig. 2.3 are produced with the following commands:

```
> hl2 = list(arcs = vstructs(dag2, arcs = TRUE),
+         lwd = 4, col = "black")
> hl3 = list(arcs = vstructs(dag3, arcs = TRUE),
+         lwd = 4, col = "black")
> graphviz.plot(dag2, highlight = hl2, layout = "fdp",
+    main = "dag2")
> graphviz.plot(dag3, highlight = hl3, layout = "fdp",
+    main = "dag3")
> graphviz.plot(cpdag(dag2), highlight = hl2,
+    layout = "fdp", main = "cpdag(dag2)")
> graphviz.plot(cpdag(dag3), highlight = hl3,
+    layout = "fdp", main = "cpdag(dag3)")
```

**bnlearn** also provides a `plot` method for `bn` objects, which is very similar to the one implemented in **deal** for `network` objects. Both methods are limited by their simple layout algorithms, which are not able to produce clear plots for large networks, and by the limited number of graphical parameters.

**pcalg** and **catnet** do not implement any native plotting function, relying instead on the functionality provided by packages **graph** and **Rgraphviz** and package **igraph** (Csardi and Nepusz, 2006), respectively. **pcalg** provides a wrapper in the form of a `plot` method for `pcAlgo` objects, while **catnet** calls it `cnPlot`. Both these approaches allow a fine control on the layout and the formatting of the plot through the options provided by the supporting packages mentioned above.

## *2.3.4 Structure Learning*

So far, we have analyzed the `marks` data set using pre-specified network structures. While this approach may be feasible in some settings, such as when expert knowledge is available, it is far more common for the network structure to be learned from the data. For this reason, we will now focus on the various options available in R for structure learning.

Consider, for instance, the network structure learned for the `marks` data with the Grow-Shrink implementation from **bnlearn** (Fig. 2.4).

```
> bn.gs = gs(marks)
> bn.gs

  Bayesian network learned via Constraint-based
  methods

  model:
   [STAT] [ANL|STAT] [ALG|ANL:STAT] [VECT|ALG]
   [MECH|VECT:ALG]
  nodes:                                    5
  arcs:                                     6
    undirected arcs:                        0
    directed arcs:                          6
  average markov blanket size:              2.40
  average neighbourhood size:               2.40
  average branching factor:                 1.20

  learning algorithm:                Grow-Shrink
  conditional independence test:
                      Pearson's Linear Correlation
  alpha threshold:                          0.05
  tests used in the learning procedure:     32
  optimized:                                TRUE
```

The default conditional independence test is the Student's $t$ test introduced in Eq. 2.13, because of its exact distribution, with a threshold of $\alpha = 0.05$ for the type I error. Note that constraint-based algorithms are largely self-correcting for multiplicity (Aliferis et al., 2010a,b); no explicit multiplicity correction such as *family-wise error rate* (FWER) or *false discovery rate* (FDR) is needed to choose a suitable threshold. Small values of $\alpha$, e.g. $\alpha \in [0.01, 0.05]$ work well for networks with up to hundreds of variables.

All the IAMB algorithms return the same network structure as `gs`, which in turn is identical to the DAG in Fig. 2.2. Even changing the conditional independence test to Fisher's $Z$ or performing a permutation test (by setting the `test` argument to `zf` or `mc-cor`, respectively) does not make any difference.

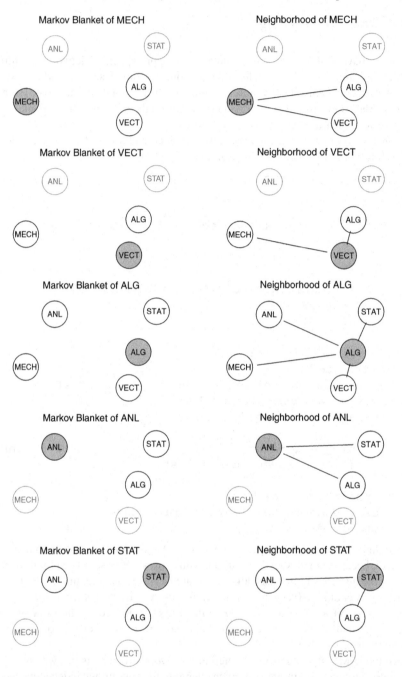

**Fig. 2.4** Markov blanket (on the *left*) and neighborhood (on the *right*) learned by the Grow-Shrink algorithm for each node (*shaded*)

```
> all.equal(bn.gs, iamb(marks))
[1] TRUE
> all.equal(bn.gs, inter.iamb(marks))
[1] TRUE
> all.equal(bn.gs, iamb(marks, test = "mc-cor"))
[1] TRUE
```

The implementation of the PC algorithm provided by the **pcalg** package, which is invoked through the pc function, produces the following output:

```
> suff.stat = list(C = cor(marks), n = nrow(marks))
> pc.fit = pc(suff.stat, indepTest = gaussCItest,
+              p = ncol(marks), alpha = 0.05)
> pc.fit
Object of class 'pcAlgo', from Call:
  skeleton(suffStat = suffStat, indepTest = indepTest,
    p = p, alpha = alpha, verbose = verbose,
    fixedGaps = fixedGaps, fixedEdges = fixedEdges,
    NAdelete = NAdelete, m.max = m.max)
A graphNEL graph with directed edges
Number of Nodes = 5
Number of Edges = 7
```

The default options are very similar to the ones used above with **bnlearn**; the gaussCItest function (which is provided by **pcalg**) implements Fisher's $Z$ test, and again $\alpha = 0.05$. We can also supply a function implementing a custom conditional independence test via the indepTest argument. Such a function must take the labels of two nodes, the set of d-separating nodes, and the suff.stat object as arguments, and return the p-value of the test.

The easiest way to compare pc.fit with bn.graph is to use the classes provided by the **graph** package, which **pcalg** uses in the pcAlgo class, and the compareGraphs function provided by **pcalg** itself.

```
> gs.graph = as.graphAM(bn.gs)
> compareGraphs(pc.fit@graph, gs.graph)
tpr fpr tdr
  1   0   1
```

As we can see, the *true positive rate* (TPR) and the *true discovery rate* (TDR) of the arcs in pc.fit are both equal to one. In other words, the proportion of arcs that are correctly identified in pc.fit with respect to gs.graph is 1, which implies that the two network structures are identical.

Considering again the algorithms implemented in **bnlearn**, we can see that hill-climbing and MMHC learn a different network structure from constraint-based algorithms. The network learned with the hill-climbing algorithm is shown below, and the steps performed by the algorithm are shown in Fig. 2.5.

```
> bn.hc = hc(marks)
```

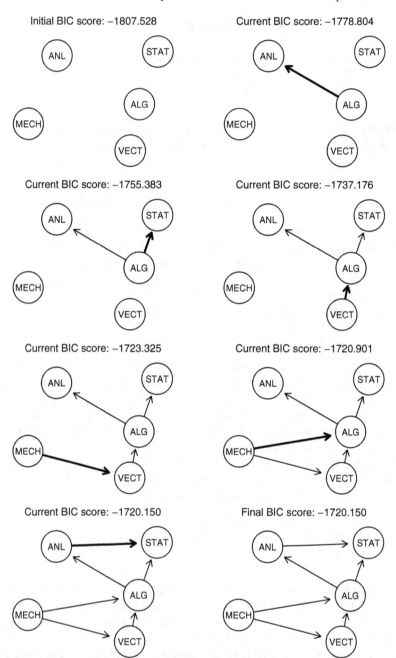

**Fig. 2.5** Operations performed by the hill-climbing algorithm (as implemented in **bnlearn**) for learning of the structure of the marks data set. The arc added in each step is highlighted with a *thicker line width*

```
> bn.hc

  Bayesian network learned via Score-based methods

  model:
   [MECH] [VECT|MECH] [ALG|MECH:VECT] [ANL|ALG]
   [STAT|ALG:ANL]
  nodes:                                        5
  arcs:                                         6
     undirected arcs:                           0
     directed arcs:                             6
  average markov blanket size:                  2.40
  average neighbourhood size:                   2.40
  average branching factor:                     1.20

  learning algorithm:
                                       Hill-Climbing
  score:
              Bayesian Information Criterion (Gaussian)
  penalization coefficient:             2.238668
  tests used in the learning procedure:  34
  optimized:                            TRUE
```

This new network fits the marks data as well as the previous one, as they have the same BIC score.

```
> score(bn.gs, data = marks, type = "bic-g")
[1] -1720.15
> score(bn.hc, data = marks, type = "bic-g")
[1] -1720.15
```

Even though the directions of the arcs are different, the arcs themselves are the same in all the Bayesian networks we learned from the data. Furthermore, they are also the same (again modulo their direction) as the ones present in the original model from Whittaker (1990), which seems to indicate that the network structure we learned is reliable.

Learning the structure of a network with **deal** requires a different workflow due to the Bayesian approach to model selection adopted by the package, even though it leads to the same result. First, we must define an empty network structure for the data and fit the prior distribution, which has the form described in Bøttcher and Dethlefsen (2003).

```
> net = network(marks)
> prior = jointprior(net, N = 5)
```

The argument N passed to jointprior is the *imaginary* or *equivalent sample size*, which expresses the weight assigned to the prior distribution as the size of an imaginary sample size supporting it. According to the experiments performed by Koller and Friedman (2009), low values result in a good balance between the smoothing effect of the uniform prior distribution and the accuracy of the model, so we set N = 5.

Once the prior distribution has been set up, we can fit the posterior distribution for net and use it as the starting point of a hill-climbing search for the network structure with the highest posterior density.

```
> net = learn(net, marks, prior)$nw
> best = autosearch(net, marks, prior)
> mstring = deal::modelstring(best$nw)
> mstring
[1]  "[MECH|ALG] [VECT|MECH:ALG] [ALG|ANL] [ANL]
[STAT|ALG:ANL]"
```

As we can see from the model string, the network returned by the heuristic search is very similar but not identical to the one returned by the implementation of hill-climbing present in **bnlearn**. This could be the result of using different parameters for the search (i.e., the BIC score instead of the posterior density, different default values for the imaginary sample size, etc.). However, we can show that the two networks have the same score and are therefore both optimal.

```
> bn.deal = model2network(mstring)
> bnlearn::score(bn.deal, marks, type = "bge",
  iss = 5)
[1] -1725.729
> bn.hc = hc(marks)
> bnlearn::score(bn.hc, marks, type = "bge", iss = 5)
[1] -1725.729
```

Note that the double-colon syntax (deal:: and bnlearn::) is required to execute the correct function, because both **bnlearn** and **deal** provide functions called modelstring.

The stability of this network structure can be confirmed by the use of a second search with random restarts, which are also implemented in **bnlearn** for hc (and controlled with the restart and perturb arguments).

```
> heuristic = heuristic(best$nw, marks, prior,
+                   restart = 2, trylist = best$trylist)
```

### 2.3.5 Parameter Learning

Once we have learned the network structure, we can estimate the parameters of the local distributions. For every package with the exception of **bnlearn** this step is executed by the same functions that learn the structure of the network. In addition, only one estimator is implemented: either a maximum likelihood estimator or a Bayesian one.

When using **bnlearn**, parameter learning is performed by the bn.fit function, which takes the network structure and the data as parameters. Since marks is a continuous data set, the parameters take the form of regression coefficients as indicated in Sect. 2.2.4. Their maximum likelihood estimates can be computed as follows. Only a single node is shown for brevity:

```
> fitted = bn.fit(bn.gs, data = marks)
> fitted$ALG


  Parameters of node ALG (Gaussian distribution)

Conditional density: ALG | ANL + STAT
Coefficients:
(Intercept)             ANL            STAT
 24.7254768       0.3482454       0.2273881
Standard deviation of the residuals: 6.791987
```

In general, we can specify which estimator will be used via the method argument, which can be set either to "mle" (for the maximum likelihood estimator) or "bayes" (for the posterior Bayesian estimator arising from a flat, non-informative prior). Only the former is available for continuous data.

Once we have a bn.fit object, we can modify any local distribution with the usual replacement operators. In the case of Gaussian Bayesian networks, the new parameters should be specified in a list containing at least a complete set of regression coefficients (coef) and the standard deviation of the residuals (sd).

```
> fitted$ALG = list(coef = c("(Intercept)" = 25,
+               "ANL" = 0.5, "STAT" = 0.25), sd = 6.5)
> fitted$ALG


  Parameters of node ALG (Gaussian distribution)

Conditional density: ALG | ANL + STAT
Coefficients:
(Intercept)             ANL            STAT
      25.00            0.50            0.25
Standard deviation of the residuals: 6.5
```

New sets of fitted values (fitted) and residuals (resid) can also be included in the assignment.

In addition, we can create a bn.fit object from scratch using the custom.fit function and specifying the parameters with same syntax used above.

```
> MECH.par = list(coef = c("(Intercept)" = -10,
+               "VECT" = 0.5, "ALG" = 0.6), sd = 13)
> VECT.par = list(coef = c("(Intercept)" = 10,
```

```
+                                      "ALG" = 1), sd = 10)
> ALG.par = list(coef = c("(Intercept)" = 25,
+                    "ANL" = 0.5, "STAT" = 0.25), sd = 6.5)
> ANL.par = list(coef = c("(Intercept)"  = 25,
+                                      "STAT" = 0.5), sd = 12)
> STAT.par = list(coef = c("(Intercept)"  = 43),
+                                                sd = 17)
> dist = list(MECH = MECH.par, VECT = VECT.par,
+              ALG = ALG.par, ANL = ANL.par,
+              STAT = STAT.par)
> fitted2 = custom.fit(bn.gs, dist = dist)
```

Note that the network structure stored in the object of class bn passed to bn.fit and custom.fit must be a DAG; any undirected arc must be either dropped (with the drop.arc function) or replaced with a directed one (with the set.arc function). As an alternative, if the network structure is a completed partially acyclic graph representing an equivalence class, we can also use the cextend function to consistently extend it to a DAG (Dor and Tarsi, 1992).

### 2.3.6 Discretization

We consider now how to discretize the marks data set while at the same time preserving the dependence structure of the data and how this transformation changes the results of Bayesian network learning. For instance, we can discretize each variable in the marks data into a dichotomic one by a median split transform (so that students with marks above the median are in one category and students below the median are in the other one). If we learn the network structure of this new data set, using the Grow-Shrink and hill-climbing algorithms as we did in Sect. 2.3.4, we get the networks shown in Fig. 2.6.

```
> dmarks = discretize(marks, breaks = 2,
+                                 method = "interval")
> bn.dgs = gs(dmarks)
> bn.dhc = hc(dmarks)
> all.equal(cpdag(bn.dgs), cpdag(bn.dhc))
[1] TRUE
```

Both networks belong again to the same equivalence class, and we can see that part of the dependence structure of the original network is still present: ALG still d-separates ANL and STAT from MECH and VECT, but the arc between ANL and STAT and the one between ALG and MECH are missing.

Since all the variables are now discrete, the parameters of the Bayesian network are the elements of the CPTs, as discussed in Sect. 2.2.4. For example, for the bn.dhc network, they can be learned and displayed as follows:

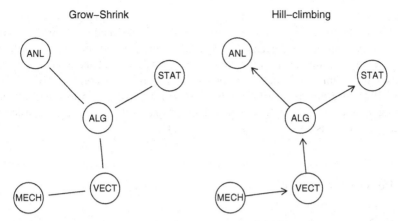

**Fig. 2.6** Bayesian networks learned from the discretized marks data using Grow-Shrink (on the *left*) and hill-climbing (on the *right*). Both belong to the same equivalence class and thus share the same CPDAG representation

```
> fitted = bn.fit(bn.dhc, data = dmarks)
> fitted$ALG

  Parameters of node ALG (multinomial distribution)

Conditional probability table:

               VECT
ALG             [8.93,45.5]  (45.5,82.1]
   [14.9,47.5]    0.5806452    0.2280702
   (47.5,80.1]    0.4193548    0.7719298
```

Now that all the variables are discrete, we can also use package **catnet**, which does not handle continuous data. Since we do not know the correct topological ordering of the nodes, we will need to call both the cnSearchSA and the cnFindBIC functions. The former performs a simulated annealing search in the space of orderings, returning the networks with the highest likelihood given their respective node orderings; the latter then returns the network with the highest BIC score among them.

```
> netlist = cnSearchSA(dmarks)
> best = cnFindBIC(netlist, nrow(dmarks))
> cnMatEdges(best)
       [,1]    [,2]
[1,]  "MECH"  "VECT"
[2,]  "VECT"  "ALG"
[3,]  "ALG"   "ANL"
[4,]  "ALG"   "STAT"
```

As we can see from the output of cnMatEdges, this learning strategy returns a network similar to bn.dgs and bn.dhc. However, if we run this example multiple times, occasionally we will get a network in which the arc between MECH and VECT is missing. This is the result of the natural sensitivity of simulated annealing to the values of its parameters, which are known to be difficult to set correctly (Bouckaert, 1995). If we use the cnSearchOrder function instead of cnSearchSA, thus limiting the search of the optimal network to the ones with the same node ordering as bn.dhc, this instability disappears completely.

## 2.4 Pearl's Causality

In Sect. 2.1, Bayesian networks have been defined in terms of conditional independence statements and probabilistic properties, without any implication that arcs should represent cause-and-effect relationships. The existence of equivalence classes of networks indistinguishable from a probabilistic point of view provides a simple proof that arc directions are not indicative of causal effects.

However, from an intuitive point of view, it can be argued that a "good" Bayesian network should represent the causal structure of the data it is describing. Such networks are usually fairly sparse, and their interpretation is at the same time clear and meaningful, as explained by Pearl (2009) in his book on causality:

> It seems that if conditional independence judgments are byproducts of stored causal relationships, then tapping and representing those relationships directly would be a more natural and more reliable way of expressing what we know or believe about the world. This is indeed the philosophy behind causal Bayesian networks.

Learning causal models, especially from observational data, presents significant challenges. In particular, three additional assumptions are needed compared to non-causal Bayesian network learning:

- Each variable $X_i \in \mathbf{X}$ is conditionally independent of its non-effects, both direct and indirect, given its direct causes. This assumption is called the *causal Markov assumption* and represents the causal interpretation of the Markov property introduced in Sect. 2.1.
- There must exist a network structure which is faithful to the dependence structure of $\mathbf{X}$.
- There must be no *latent variables* (unobserved variables influencing the variables in the network) acting as *confounding factors*. Such variables may induce spurious correlations between the observed variables, thus introducing bias in the causal network. Even though this is often listed as a separate assumption, it is really a corollary of the first two: the presence of unobserved variables violates the faithfulness assumption (because the network structure does not include them) and possibly the causal Markov property (if an arc is wrongly added between the observed variables due to the influence of the latent one).

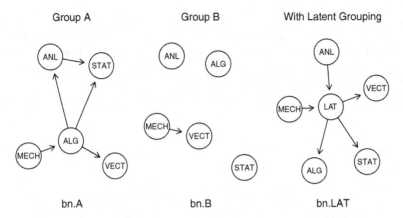

**Fig. 2.7** Bayesian networks learned from each group of students (*left* and *center*) and the network learned from the whole discretized data set after the inclusion of the latent variable LAT

These assumptions are difficult to verify in real-world settings, as the set of the potential confounding factors is not usually known. At best we can address this issue, along with selection bias, by implementing a carefully planned experimental design.

Furthermore, even when dealing with interventional data collected from a controlled experiment (where we can set the value of at least some variables and observe the resulting changes), there are usually multiple equivalent network structures that represent reasonable causal models. Many arcs may not have a definite direction, resulting in substantially different networks. When the sample size is small, there may also be several non-equivalent networks fitting the data equally well. Therefore, in general we are not able to identify a single, "best," causal network but rather a small set of likely causal networks that fit our knowledge of the data.

An example of the bias introduced by the presence of a latent variable was illustrated by Edwards (2000) using the `marks` data. He noted that if we assume that the students belong to two different groups (which we will call A and B) and assign each student to one of them using the EM algorithm (MacLachlan and Krishnan, 2008), each group identifies a different set of relationships between the five topics.

```
> latent = factor(c(rep("A", 44), "B",
+                    rep("A", 7), rep("B", 36)))
> bn.A = hc(marks[latent == "A", ])
> bn.B = hc(marks[latent == "B", ])
> modelstring(bn.A)
[1]  "[MECH] [ALG|MECH] [VECT|ALG] [ANL|ALG]
[STAT|ALG:ANL] "
> modelstring(bn.B)
[1]  "[MECH] [ALG] [ANL] [STAT] [VECT|MECH] "
```

The EM algorithm assigned the first 52 students (with the exception of number 45) to belong to group A and the remainder to group B. If we consider once more the discretized marks dmarks we created in Sect. 2.2.6 and include the latent variable when learning the structure of the network, we obtain a network structure completely different from the ones in Fig. 2.6.

```
> bn.LAT = hc(cbind(dmarks, LAT = latent))
> bn.LAT


Bayesian network learned via Score-based methods

model:
  [MECH] [ANL] [LAT|MECH:ANL] [VECT|LAT] [ALG|LAT]
  [STAT|LAT]
nodes:                                      6
arcs:                                       5
   undirected arcs:                         0
   directed arcs:                           5
average markov blanket size:             2.00
average neighbourhood size:              1.67
average branching factor:                0.83

learning algorithm:
                                    Hill-Climbing
score:
                       Bayesian Information Criterion
penalization coefficient:            2.238668
tests used in the learning procedure:  40
optimized:                             TRUE
```

The three network structures learned above are shown in Fig. 2.7; the one including the latent variable, bn.LAT, agrees with the network structure reported in Edwards (2000). We can clearly see that any causal relationship we could infer without taking LAT into account would be potentially spurious. In fact, we could even question the assumption that the data are a random sample from a single population and have not been manipulated in some way beforehand.

## 2.5 Applications to Gene Expression Profiles

Static Bayesian networks provide a versatile tool for the analysis of many kinds of biological data, including (but not limited to) single-nucleotide polymorphism (SNP) data and gene expression profiles. Following the work of Friedman et al. (2000), the expression level or the allele frequency of each gene is associated with one node. In addition, we can include in the network additional nodes denoting other

attributes that affect the system, such as experimental conditions, temporal indicators, and exogenous cellular conditions. As a result, we can model simultaneously
the biological mechanisms we are interested in and the external conditions influencing them in a single, comprehensive network.

## 2.5.1 Model Averaging

Consider, for example, the protein signaling data studied in Sachs et al. (2005). The
data consist in the simultaneous measurements of 11 phosphorylated proteins and
phospholypids derived from thousands of individual primary immune system cells,
subjected to both general and specific molecular interventions. The former ensure
that the relevant signaling pathways are active, while the latter make causal inference possible by elucidating arc directions through stimolatory cues and inhibitory
interventions.

The analysis performed in Sachs et al. (2005) can be summarized as follows:

1. Outliers were removed and the data were discretized using the approach described in Hartemink (2001), because the distributional assumptions required by
   Gaussian Bayesian networks were unlikely to hold.
2. Structure learning was repeated several times. In this way, a larger number of network structures were explored in an effort to reduce the impact of locally optimal
   (but globally suboptimal) networks on learning and subsequent inference.
3. The networks learned in the previous step were averaged to produce a more
   robust model. This practice, known as *model averaging* (Claeskens and Hjort,
   2008), is known to result in a better predictive performance than choosing a single, high-scoring network. The averaged network structure was created using
   the arcs present in at least 85 % of the networks. This proportion measures the
   *strength* of each arc and provides the means to establish its *significance* given a
   *threshold* (85 % in this case).
4. The validity of the averaged network was evaluated using connections well-
   established in literature as a reference.

All these steps can be performed using the **bnlearn** package, and some of the other
packages covered in Sect. 2.3.1 can also be used by integrating missing functionality. For the moment, we will consider only the data manipulated with general
interventions (i.e., the observational data); we will investigate the complete data set
(i.e., both the observational and the interventional data) in Sect. 2.5.3.

First of all, we will discretize the data with the discretize function, which
implements some common discretization methods including Hartemink's.

```
> library(bnlearn)
> sachs = read.table("sachs.data.txt", header = TRUE)
> dsachs = discretize(sachs, method = "hartemink",
+             breaks = 3, ibreaks = 60,
+             idisc = "quantile")
```

The relevant arguments are idisc and ibreaks, which control how the data are initially discretized, and breaks, which specifies the number of levels of each discretized variable. Choosing good values for these arguments is a trade-off between quality and speed; high values of idisc preserve the characteristics of the original data to a greater extent, whereas smaller values result in much smaller memory usage and shorter running times.

Each variable in the dsachs data frame is a factor with three levels, corresponding roughly to low, average, and high expression. Now that the data are ready for the analysis, we can apply bootstrap resampling to dsachs to learn a set of 500 network structures to use for model averaging.

```
> boot = boot.strength(data = dsachs, R = 500,
+              algorithm = "hc",
+              algorithm.args = list(score = "bde",
+                                    iss = 10))
> boot[(boot$strength > 0.85) &
+                     (boot$direction >= 0.5), ]
          from        to strength direction
1         praf      pmek    1.000     0.573
23        plcg      PIP2    1.000     0.624
24        plcg      PIP3    1.000     0.994
34        PIP2      PIP3    1.000     0.997
56      p44.42  pakts473    1.000     0.616
57      p44.42       PKA    0.998     1.000
67    pakts473       PKA    1.000     0.995
89         PKC       P38    1.000     0.530
90         PKC      pjnk    1.000     0.982
100        P38      pjnk    0.962     1.000
```

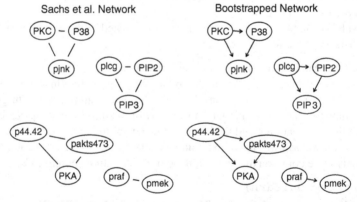

**Fig. 2.8** Averaged network learned from the observational data studied in Sachs et al. (2005) (on the *left*) and the network learned with bootstrap resampling from the same data (on the *right*)

The code above implements a setup similar to Sachs et al. (2005). We learn a network structure from each bootstrap sample with a hill-climbing search and a BDe posterior density with a very low imaginary sample size. Arcs are considered significant if they appear in at least 85 % of the networks and in the direction that appears most frequently. Interestingly, lowering the threshold to 50 % does not change the results of the analysis, which seems to indicate that its value is not critical in this case.

Having computed the significance for all possible arcs, we can now build the averaged network structure as follows:

```
> avg.boot = averaged.network(boot, threshold = 0.85)
```

As shown in Fig. 2.8, the averaged network contains the same arcs as the network learned from the observational data in Sachs et al. (2005). The only difference is that since the probability of the presence of each arc and of its possible directions are computed separately in boot.strength, we can also determine which direction has better support from the discretized data.

As an alternative, we can average the results of several hill-climbing searches, each starting from a different network. Such networks can be generated randomly from a uniform distribution over the space of connected graphs with the algorithm proposed by Ide and Cozman (2002) and implemented in random.graph. This ensures both that no systematic bias is introduced in the learned networks and that the search space is covered more thoroughly.

```
> nodes = names(dsachs)
> start = random.graph(nodes = nodes,
+             method = "ic-dag", num = 500)
> netlist = lapply(start, function(net) {
+    hc(dsachs, score = "bde", iss = 10, start = net)
+ })
> rnd = custom.strength(netlist, nodes = nodes)

> rnd[(rnd$strength > 0.85) &
+             (rnd$direction >= 0.5), ]
          from       to strength direction
11        pmek     praf        1     0.518
23        plcg     PIP2        1     0.748
43        PIP3     plcg        1     0.542
44        PIP3     PIP2        1     0.790
56      p44.42 pakts473        1     0.682
57      p44.42      PKA        1     0.704
67     pakts473     PKA        1     0.522
89         PKC      P38        1     0.540
90         PKC     pjnk        1     0.945
100        P38     pjnk        1     0.905
> avg.start = averaged.network(rnd, threshold = 0.85)
```

The arcs identified as significant with this approach are the same as in `avg.boot` (even though some of them are reversed), thus confirming the stability of the averaged network obtained from bootstrap resampling. A comparison of the equivalence classes of `avg.boot` and `avg.start` suggests that the two networks are equivalent.

```
> all.equal(cpdag(avg.boot), cpdag(avg.start))
[1] TRUE
```

Furthermore, note that averaged networks, like the networks they are computed from, are not necessarily completely directed. In that case, it is not possible to compute their score directly. However, we can identify the equivalence class the averaged network belongs to (with `cpdag`) and then select a DAG within that equivalence class (with `cextend`).

```
> score(cextend(cpdag(avg.start)), dsachs,
+    type = "bde", iss = 10)
[1] -8498.877
```

Since all networks in the same equivalence class have the same score (for score equivalent functions), the value returned by `score` is a correct estimate for the original, partially directed network.

We can also compute such averaged network structures from bootstrap samples using the algorithms implemented in **catnet**, **deal**, and **pcalg**. For this purpose, bnlearn provides a `custom.strength` function that requires only a list of arc sets as argument, thus facilitating interoperability.

For example, we can replace the hill-climbing search used above with the simulated annealing search implemented in **catnet** as follows:

```
> library(catnet)
> netlist = vector(500, mode = "list")
> ndata = nrow(dsachs)

> netlist = lapply(netlist, function(net) {
+    boot = dsachs[sample(ndata, replace = TRUE), ]
+    nets = cnSearchOrder(boot)
+    best = cnFindBIC(nets, ndata)
+    cnMatEdges(best)
+ })
> sa = custom.strength(netlist, nodes = nodes)
> sa[(sa$strength > 0.85) &
+                       (sa$direction >= 0.5), ]
        from      to strength direction
1       praf    pmek     1.00       0.5
11      pmek    praf     1.00       0.5
23      plcg    PIP2     0.99       0.5
33      PIP2    plcg     0.99       0.5
34      PIP2    PIP3     1.00       0.5
```

```
44         PIP3         PIP2      1.00          0.5
56       p44.42     pakts473      1.00          0.5
66     pakts473       p44.42      1.00          0.5
67     pakts473          PKA      1.00          0.5
77          PKA     pakts473      1.00          0.5
89          PKC          P38      1.00          0.5
90          PKC         pjnk      1.00          0.5
99          P38          PKC      1.00          0.5
109        pjnk          PKC      1.00          0.5
> avg.catnet = averaged.network(sa, threshold = 0.85)
```

Again, `avg.catnet` presents some small differences from both `avg.boot` and `avg.start`. Such differences can be attributed to the different scores and structure learning algorithms used to build the sets of high-scoring networks. In particular, it is very common for arc directions to change between different learning algorithms as a result of score equivalence.

## 2.5.2 Choosing the Significance Threshold

The value of the threshold beyond which an arc is considered significant, which is often called the *significance threshold*, does not seem to have a huge influence on the analysis of the data analyzed in Sachs et al. (2005). In fact, any value between 0.5 and 0.85 yields exactly the same results. So, for instance,

```
> all.equal(averaged.network(boot, threshold = 0.50),
+           averaged.network(boot, threshold = 0.70))
[1] TRUE
```

The same holds for `avg.catnet` and `avg.start`. However, this is often not the case. Therefore, it is important to use a statistically motivated algorithm for choosing a suitable threshold instead of relying on ad hoc values.

A solution to this problem is presented in Scutari and Nagarajan (2012) and implemented in **bnlearn** as the default value for the `threshold` argument in `averaged.network`.

```
> averaged.network(boot)

  Random/Generated Bayesian network

  model:
   [praf] [plcg] [p44.42] [PKC] [pmek|praf] [PIP2|plcg]
   [pakts473|p44.42] [P38|PKC] [PIP3|plcg:PIP2]
   [PKA|p44.42:pakts473] [pjnk|PKC:P38]
  nodes:                                           11
```

```
arcs:                                    10
   undirected arcs:                      0
   directed arcs:                        10
average markov blanket size:             1.82
average neighbourhood size:              1.82
average branching factor:                0.91

generation algorithm:
                                    Model Averaging
   significance threshold:              0.374
```

The value of the threshold is computed as follows. If we denote the arc strengths stored in boot as $\hat{\mathbf{p}} = \{\hat{p}_i, i = 1, \ldots, k\}$ and $\hat{\mathbf{p}}_{(\cdot)}$ is

$$\hat{\mathbf{p}}_{(\cdot)} = \{0 \leqslant \hat{p}_{(1)} \leqslant \hat{p}_{(2)} \leqslant \ldots \leqslant \hat{p}_{(k)} \leqslant 1\}, \qquad (2.16)$$

then we can define the corresponding arc strengths in the (unknown) averaged network $G = (\mathbf{V}, A_0)$ as

$$\tilde{p}_{(i)} = \begin{cases} 1 & \text{if } a_{(i)} \in A_0 \\ 0 & \text{otherwise} \end{cases}, \qquad (2.17)$$

that is, the set of strengths that characterizes any arc as either significant or non-significant without any uncertainty. In other words,

$$\tilde{\mathbf{p}}_{(\cdot)} = \{0, \ldots, 0, 1, \ldots, 1\}. \qquad (2.18)$$

The proportion $t$ of elements of $\tilde{\mathbf{p}}_{(\cdot)}$ that are equal to 1 determines the number of arcs in the averaged network and is a function of the significance threshold we want to estimate. One way to do that is to find the value $\hat{t}$ that minimizes the $L_1$ norm

$$L_1\left(t; \hat{\mathbf{p}}_{(\cdot)}\right) = \int \left| F_{\hat{\mathbf{p}}_{(\cdot)}}(x) - F_{\tilde{\mathbf{p}}_{(\cdot)}}(x; t) \right| dx \qquad (2.19)$$

between the cumulative distribution functions of $\tilde{\mathbf{p}}_{(\cdot)}$ and $\hat{\mathbf{p}}_{(\cdot)}$ and then to include every arc that satisfies

$$a_{(i)} \in A_0 \iff \hat{p}_{(i)} > F_{\hat{\mathbf{p}}_{(\cdot)}}^{-1}(\hat{t}) \qquad (2.20)$$

in the average network. This amounts to finding the averaged network whose arc set is "closest" to the arc strength computed from the data, with $F_{\hat{\mathbf{p}}_{(\cdot)}}^{-1}(\hat{t})$ acting as the significance threshold.

For the dsachs data, the estimated value for the threshold is 0.374; so, any arc with a strength value strictly greater than that is considered significant. Again, the resulting averaged network is the same as the one obtained with the original threshold in Sachs et al. (2005).

```
> all.equal(avg.boot, averaged.network(boot))
[1] TRUE
```

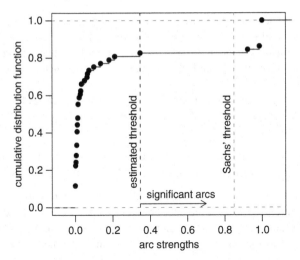

**Fig. 2.9** Cumulative distribution function of the arc strength values computed with bootstrap re-sampling from dsachs. The *vertical dashed lines* correspond to the estimated (*black*) and ad hoc (*grey*) significance thresholds

The reason for the insensitivity of the averaged network to the value of the threshold is apparent from the plot of $F_{\hat{p}_{(\cdot)}}$ in Fig. 2.9: arcs which are well supported by the data are clearly separated from the ones that are not. Since the lowest strength coefficient in the first set is 0.962 and the highest one in the second set is 0.347 (i.e., the estimated threshold), any threshold that falls between those two values results in the same averaged network.

### 2.5.3 Handling Interventional Data

Usually, all the observations in a sample are collected under the same general conditions. This is true both for observational data, in which treatment allocation is outside the control of the investigator, and for experimental data, which are collected from randomized controlled trials. As a result, the sample can be modeled with a single Bayesian network, because all the observations follow the same probability distribution.

However, this is not the case when several samples resulting from different experiments are analyzed together with a single, encompassing model. Such an approach is called *meta-analysis* (see Kulinskaya et al., 2008, for a gentle introduction). First, environmental conditions and other exogenous factors may differ between those experiments. Furthermore, the experiments may be different in themselves; for example, they may explore different treatment regimes or target different populations.

This is the case with the protein signaling data analyzed in Sachs et al. (2005). In addition to the data set we have analyzed so far, which is subject only to a general stimulus meant to activate the desired paths, nine other data sets subject to different stimulatory cues and inhibitory interventions are used to elucidate the direction of the causal relationships in the network. Such data are often called *interventional*, because the values of specific variables in the model are set by an external intervention of the investigator.

Overall, the ten data sets contain 5,400 observations; in addition to the 11 signaling levels analyzed above, the protein which is activated or inhibited (INT) is recorded for each sample.

```
> isachs = read.table("sachs.interventional.txt",
+              header = TRUE, colClasses = "factor")
```

One intuitive way to model these data sets with a single, encompassing model is to include the intervention INT in the network and to make all variables depend on it. This can be achieved with a whitelist containing all possible arcs from INT to the other nodes, thus forcing such arcs to be present in the learned network structure.

```
> wh = matrix(c(rep("INT", 11), names(isachs)[1:11]),
+         ncol = 2)
> bn.wh = tabu(isachs, whitelist = wh, score = "bde",
+              iss = 10, tabu = 50)
```

Using tabu search instead of hill-climbing improves the stability of the score-based search; once a locally optimum network is found, tabu search performs an additional 50 iterations (as specified by the tabu argument) to ensure that no other (and potentially better) local optimum is found.

We can also let the structure learning algorithm decide which arcs connecting INT to the other nodes should be included in the network. To this end, we can use the tiers2blacklist function to blacklist all arcs toward INT, thus ensuring that only outgoing arcs will be included in the network. In the general case, tiers2blacklist builds a blacklist such that all arcs going from a node in a particular element of the nodes argument to a node in one of the previous elements are blacklisted.

```
> tiers = list("INT", names(isachs)[1:11])
> bl = tiers2blacklist(nodes = tiers)
> bn.tiers = tabu(isachs, blacklist = bl,
+              score = "bde", iss = 10, tabu = 50)
```

The networks learned with these two approaches are shown in Fig. 2.10. Some of the structural features detected in Sachs et al. (2005) are present in both bn.wh and bn.tiers. For example, the interplay between plcg, PIP2, and PIP3 and between PKC, P38, and pjnk are both correctly modeled. The lack of any direct intervention on PIP2 is also correctly modeled in bn.tiers. The most noticeable feature missing from both networks is the pathway linking praf to pakt473 through pmek and p44.42.

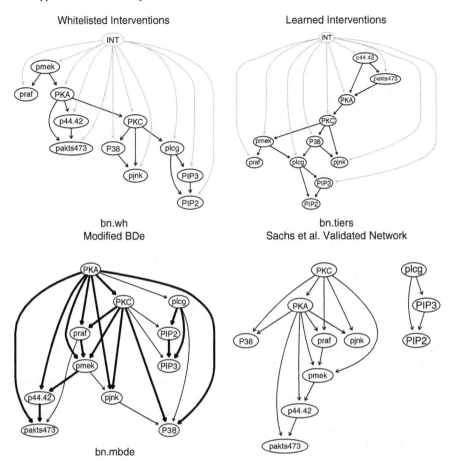

**Fig. 2.10** Bayesian networks learned from isachs. The first two networks (bn.wh on the *top left*, bn.tiers on the *top right*) have been learned by including INT and adding arcs to model stimulatory cues and inhibitory interventions. The third network (bn.mbde, on the *bottom left*) has been learned with model averaging and the mbde score; arcs highlighted with a *thicker line width* make up the validated Bayesian network (*bottom right*) from Sachs et al. (2005)

The approach used in Sachs et al. (2005) yields much better results. Instead of including the interventions in the network as an additional node, Sachs et al. (2005) used a modified BDe score (labeled "mbde" in **bnlearn**) incorporating the effects of the interventions into the score components associated with each node (Cooper and Yoo, 1995).

Since the value of INT identifies which node is subject to either a stimulatory cue or an inhibitory intervention for each observation, we can easily construct a named list of which observations are manipulated for each node.

```
> INT = sapply(1:11, function(x) {
+                              which(isachs$INT == x) })
> nodes = names(isachs)[1:11]
> names(INT) = nodes
```

Given such a list, we can then pass it to tabu as an additional argument for mbde. In addition, we can combine the use of mbde with model averaging and random starting points as discussed in Sect. 2.5.1. To improve the stability of the averaged network, we generate the set of the starting networks for the tabu searches using the algorithm from Melançon et al. (2001), which is not limited to connected networks as the one from Ide and Cozman (2002). In addition, we actually use only one generated network every 100 to obtain a more diverse set.

```
> start = random.graph(nodes = nodes,
+    method = "melancon", num = 500, burn.in = 10^5,
+    every = 100)
> netlist = lapply(start, function(net) {
+    tabu(isachs[, 1:11], score = "mbde", exp = INT,
+        iss = 10, start = net, tabu = 50) })
> arcs = custom.strength(netlist, nodes = nodes)
> bn.mbde = averaged.network(arcs, threshold = 0.85)
```

As we can see from Fig. 2.10, bn.mbde is much closer to the validated network from Sachs et al. (2005) than any of the other networks learned in this section. All the arcs from the validated network are correctly learned, even though a few are reversed. The arcs from bn.mbde that are not present in the validated networks were identified in the original paper and discarded due to their comparatively low strength; this may imply that the simulated annealing algorithm used in Sachs et al. (2005) performs better on this data set than tabu search.

## Exercises

**2.1.** Consider the asia synthetic data set from Lauritzen and Spiegelhalter (1988), which describes the diagnosis of a patient at a chest clinic who has just come back from a trip to Asia and is showing dyspnea.

(a) Load the data set from the **bnlearn** package and investigate its characteristics using the exploratory analysis techniques covered in Chap. 1.
(b) Create a bn object with the network structure described in the manual page of asia.
(c) Derive the skeleton, the moral graph, and the CPDAG representing the equivalence class of the network. Plot them using graphviz.plot.
(d) Identify the parents, the children, the neighbors, and the Markov blanket of each node.

**2.2.** Using the network structures created in Exercise 2.1 for the `asia` data set, produce the following plots with `graphviz.plot`:

(a) A plot of the CPDAG of the equivalence class in which the arcs belonging to a v-structure are highlighted (either with a different color or using a thicker line width).
(b) Fill the nodes with different colors according to their role in the diagnostic process: causes ("visit to Asia" and "smoking"), effects ("Tuberculosis," "lung cancer," and "bronchitis") and the diagnosis proper ("chest X-ray," "dyspnea," and "either tuberculosis or lung cancer/bronchitis").
(c) Explore different layouts by changing the `layout` and `shape` arguments.

**2.3.** Consider the `marks` data set analyzed in Sect. 2.3.

(a) Discretize the data using a quantile transform and different numbers of intervals (say, from 2 to 5). How does the network structure learned from the resulting data sets change as the number of intervals increases?
(b) Repeat the discretization using interval discretization using up to five intervals, and compare the resulting networks with the ones obtained previously with quantile discretization.
(c) Does Hartemink's discretization algorithm perform better than either quantile or interval discretization? How does its behavior depend on the number of initial breaks?

**2.4.** The ALARM network (Beinlich et al., 1989) is a Bayesian network designed to provide an alarm message system for patients hospitalized in intensive care units (ICU). Since ALARM is commonly used as a benchmark in literature, a synthetic data set of 5,000 observations generated from this network is available from **bnlearn** as `alarm`.

(a) Create a bn object for the "true" structure of the network using the model string provided in its manual page.
(b) Compare the networks learned with different constraint-based algorithms with the true one, both in terms of structural differences and using either BIC or BDe.
(c) The overall performance of constraint-based algorithms suggests that the asymptotic $\chi^2$ conditional independence tests may not be appropriate for analyzing `alarm`. Are permutation or shrinkage tests better choices?
(d) How are the above learning strategies affected by changes to `alpha`?

**2.5.** Consider again the `alarm` network used in Exercise 2.4.

(a) Learn its structure with hill-climbing and tabu search, using the posterior density BDe as a score function. How does the network structure change with the imaginary sample size `iss`?
(b) Does the length of the tabu list have a significant impact on the network structures learned with `tabu`?
(c) How does the BIC score compare with BDe at different sample sizes in terms of structure and score of the learned network?

**2.6.** Consider the observational data set from Sachs et al. (2005) used in Sect. 2.5.1 (the original data set, not the discretized one).

(a) Evaluate the networks learned by hill-climbing with BIC and BGe using cross-validation and the log-likelihood loss function.
(b) Use bootstrap resampling to evaluate the distribution of the number of arcs present in each of the networks learned in the previous point. Do they differ significantly?
(c) Compute the averaged network structure for sachs using hill-climbing with BGe and different imaginary sample sizes. How does the value of the significance threshold change as iss increases?

# Chapter 3
# Bayesian Networks in the Presence of Temporal Information

**Abstract** Real-world entities comprising a complex system evolve as a function of time and respond to external perturbations. Dynamic Bayesian networks extend the fundamental ideas behind static Bayesian networks to model associations arising from the temporal dynamics between the entities of interest. This has to be contrasted with static Bayesian networks, which model the network structure from multiple independent realizations of the entities of a snapshot of the process. More importantly, incorporating the temporal signatures is useful in capturing possible feedback loops that are implicitly disregarded in the case of static Bayesian networks. Since feedback loops are ubiquitous in biological pathways, dynamic Bayesian network modeling is expected to result in better representations of such pathways.

In this chapter, we will introduce basic definitions and models for modeling associations from multivariate linear time series using dynamic Bayesian networks. Applications include modeling gene networks from expression data. Two broad classes of multivariate time series are considered: those whose statistical properties are invariant as a function of time and those whose properties do show change of time.

## 3.1 Time Series and Vector Auto-Regressive Processes

### 3.1.1 Univariate Time Series

A univariate time series is a sequence of random variables

$$\{X(t)\} = \{\ldots, X(t-1), X(t), X(t+1), \ldots\} \qquad (3.1)$$

measured at successive time points, usually spaced at uniform time intervals. A univariate time series $\{X(t)\}$ is said to be *second order* or *covariance stationary*

R. Nagarajan et al., *Bayesian Networks in R: with Applications in Systems Biology*,
Use R! 48, DOI 10.1007/978-1-4614-6446-4_3,
© Springer Science+Business Media New York 2013

if the first two moments, i.e., the mean $E(X(t))$ and covariance $COV(X(t))$, are invariant as a function of time:

$$\forall t, E(X(t)) = \mu, \tag{3.2}$$
$$\forall t, i, \ COV(X(t), X(t-i)) = E((X(t) - \mu)(X(t-i) - \mu)) = \gamma_i. \tag{3.3}$$

In other words, the first two moments of covariance-stationary time series are invariant over time. In this section, *stationary time series* implicitly refer to covariance-stationary time series.

Stationary univariate time series $X(t)$ are often modeled as auto-regressive processes, where the value at a given time $t$ is given as a linear combination of those at earlier time points, $X(t-i), i = 1, \ldots, p$:

$$\forall t \geqslant p, \quad X(t) = a_1 X(t-1) + \cdots + a_i X(t-i) + \cdots + a_p X(t-p) + b + \varepsilon(t) \tag{3.4}$$

where

- $X(t)$ is the random variable observed at time $t$;
- $p$ is the *lag* or *order* of the time series;
- $a_i \in \mathbb{R}$, $i = 1, \ldots, p$, are the coefficients associated with the random variables observed at the previous $p$ time points, i.e., $t-1, t-2, \ldots, t-p$;
- $b \in \mathbb{R}$ is the baseline measurement, i.e., the intercept;
- $\varepsilon(t)$ is a Gaussian white noise, i.e., $\varepsilon(t) \sim N(0, \sigma^2)$.

### 3.1.2 Multivariate Time Series

Multivariate time series (MTS) are sequences of multivariate random variables measured at successive time points. MTS data are commonly encountered in real-world settings where the objective is to understand the associations between multiple entities from their temporal signatures. An example of MTS from Smith et al. (2004), representing the expression profiles of a set of genes, is shown in Fig. 3.1.

Multivariate time series are commonly modeled as *vector auto-regressive* (VAR) process. A VAR process is essentially a multivariate extension of an auto-regressive process. A vector auto-regressive process $VAR(p)$ of order $p$, the variables observed at any time $t \geqslant p$ are assumed to satisfy

$$X(t) = A_1 X(t-1) + \cdots + A_i X(t-i) + \cdots + A_p X(t-p) + B + \varepsilon(t) \tag{3.5}$$

where

- $X(t) = (X_i(t))$, $i = 1, \ldots, k$, is the vector of $k$ variables observed at time $t$;
- $A_i$, $i = 1, \ldots, p$ are matrices of coefficients of size $k \times k$;
- $B$ is a vector of size $k$ representing the baseline measurement for each variable;

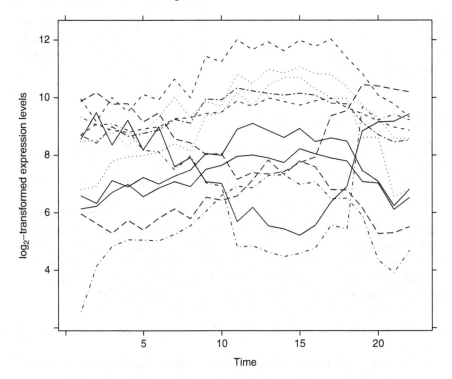

**Fig. 3.1** *Arabidopsis thaliana* gene expression time series from Smith et al. (2004)

- $\varepsilon(t)$ is a white noise vector of size $k$, with zero mean $(\mathrm{E}\,(\varepsilon(t)) = 0)$ and time-invariant positive definite covariance matrix $(\mathrm{COV}\,(\varepsilon(t)) = \Sigma)$.

Similar to an auto-regressive process, a VAR($p$) of order $p$ assumes a linear correlation structure between the $k$ variables observed at time points $t$ and the $k$ variables observed at the $p$ previous time points.

### 3.1.2.1 Covariance Stationarity of a VAR Process

A VAR($p$) process can be written as a VAR(1) process via its *companion form*,

$$Y(t) = AY(t-1) + \upsilon(t), \tag{3.6}$$

with

$$Y(t) = \begin{bmatrix} Y(t) \\ \cdot \\ \cdot \\ \cdot \\ Y(t-p+1) \end{bmatrix}, \quad A = \begin{bmatrix} A_1 & A_2 & \dots & A_{p-1} & A_p \\ I & 0 & \dots & 0 & 0 \\ 0 & I & \dots & 0 & 0 \\ \vdots & \vdots & \ddots & \vdots & \vdots \\ 0 & 0 & \dots & I & 0 \end{bmatrix}, \quad \upsilon(t) = \begin{bmatrix} \varepsilon(t) \\ 0 \\ \vdots \\ 0 \end{bmatrix}. \tag{3.7}$$

In the above, $Y(t)$ and $v(t)$ are $(kp \times 1)$ vectors, $A$ is a $(kp \times kp)$ companion matrix, and $I$ is a $(p \times p)$ identity matrix. The VAR($p$) process is said to be stationary (covariance-stationary) if the absolute values of the eigenvalues of the companion matrix $A$ are lesser than 1.

### 3.1.2.2 Lag Order of a VAR Process

The literature contains numerous discussions on how to select a suitable lag order for a covariance stationary VAR process. Among classic procedures, there are information criterion such as AIC and BIC, also known as the *Schwarz criterion* (Lütkepohl 2005).

Information criteria are statistics that measure the distance between observations and model classes. If the value of the information criterion is small, the distance is small and the model class contains a model that fits the data well. Typical criteria, such as AIC or BIC, consist of two additive parts. The first is a naïve goodness-of-fit measure, and the second is a penalty term that increases with the model's complexity. The most popular information criteria is the AIC due to Akaike:

$$AIC(m) = log|\hat{\Sigma}(m)| + \frac{2}{n}m, \tag{3.8}$$

where $m$ denotes the number of free parameters in the model and $\hat{\Sigma}(m)$ denotes the maximum likelihood estimate of the error covariance matrix. In principle, a VAR($p$) has $m = k^2p + k + k(k+1)/2$ free parameters. Since we are only interested in finding the optimal $p$, we can assume $k$ (the number of variables at each time point) is constant, discard the intercept, and not impose any constraint on the error covariance matrix. Therefore, for a VAR($p$) process, AIC is defined as

$$AIC(p) = log|\hat{\Sigma}(p)| + \frac{2pk^2}{n}. \tag{3.9}$$

An important property of AIC is its ability to select models with strong predictive power. Some authors also suggest that AIC can select good models even for small samples, possibly through the use of a second-order correction called AICc.

BIC is also commonly used because it is consistent, that is, the selected $\hat{p}$ will be the true $p$ with probability one as $n$ tends toward infinity. For a VAR($p$) process, BIC is defined as

$$BIC(p) = log|\hat{\Sigma}(p)| + \frac{pk^2log(n)}{n}. \tag{3.10}$$

### 3.1.2.3 Tests for Multivariate Normality in VARs

When using a statistical model on real-world data, it is important to check that the assumptions of the model are satisfied. In the case of VAR processes, one of those assumptions is the normality of the residuals, which can be assessed with many different tests for univariate and multivariate time series. Some examples are presented in Jarque and Bera (1980), Bera and Jarque (1981), Jarque and Bera (1987), and Lütkepohl (2005).

One such test is the Jarque–Bera normality test, a goodness-of-fit test on whether sample data have skewness and kurtosis of a normal distribution (which are both equal to zero). The test statistic is defined as

$$JB = \frac{n}{6}\left[S^2 + \frac{1}{4}(K-3)^2\right], \tag{3.11}$$

where

$$S = \frac{\hat{\mu}_3}{\hat{\sigma}^3} \quad \text{and} \quad K = \frac{\hat{\mu}_4}{\hat{\sigma}^4} \tag{3.12}$$

are the sample skewness and kurtosis; $\hat{\mu}_3$ and $\hat{\mu}_4$ are the estimates of third and fourth central moments; and $\hat{\sigma}$ is the estimate of the standard deviation. If the data come from a normal distribution, the Jarque–Bera statistic has an asymptotic $\chi^2$ distribution with $2k$ degrees of freedom.

#### 3.1.2.4  Test for Serial Correlation (Portmanteau Test)

Other diagnostic tests may be useful in analyzing multivariate time series, for instance, testing for the absence of autocorrelation, heteroscedasticity, or non-normality in $\varepsilon(t)$. The Portmanteau test and the Breusch–Godfrey serial correlation Lagrange multiplier test allow to analyze the lack of serial correlation in the residuals of a VAR($p$). Heteroscedasticity can be studied via the univariate and multivariate auto-regressive conditionally heteroscedastic Lagrange multiplier (ARCH-LM) tests for a VAR($p$) process (see Engle 1982; Hamilton 1994; Lütkepohl 2005).

These tests are implemented in the **vars** package, which will be introduced in Sect. 3.5.1.

## 3.2  Dynamic Bayesian Networks: Essential Definitions and Properties

### 3.2.1  Definitions

Unlike static Bayesian networks, each variable in a dynamic Bayesian network is represented by several nodes across time points. In essence, a dynamic Bayesian network (Fig. 3.2c) is obtained by unfolding in time an interaction graph (Fig. 3.2a). This aspect is especially useful in accommodating possible loops and feedback in the network topology. Setting the arc directions across time also guarantees the acyclicity of the graph, which is required by definition for a Bayesian network. In the resulting directed acyclic graph, an arc is drawn between two variables at successive time points, for example, from $X_1(t-1)$ to $X_2(t)$ in Fig. 3.2c, whenever these two variables are conditionally dependent given the remaining variables in the past time points. This condition provides an extension of the properties introduced in Sect. 2.1 for static Bayesian networks and, more in general, of the graphical modeling theory from Lauritzen (1996) to temporal data.

In the last decade, various network representations based on different probabilistic models have been proposed in literature: discrete models (Ong et al. 2002; Zou and Conzen 2005), VAR processes (Opgen-Rhein and Strimmer 2007), state-space or hidden Markov models (Perrin et al. 2003; Wu et al. 2004; Rangel et al. 2004; Beal et al. 2005), and nonparametric additive regression models (Imoto et al. 2002; Imoto et al. 2003; Kim et al. 2004; Sugimoto and Iba 2004). We refer the reader to Kim et al. (2003) for a comprehensive review of these models. To summarize, we present in the following a set of sufficient conditions for a model to be represented as a dynamic Bayesian network; a detailed treatment of these results is provided in Lèbre (2009).

Consider a dynamic Bayesian network with a directed acyclic graph $G$ (e.g., Fig. 3.2c) which describes a discrete-time stochastic process $\mathbf{X} = \{X_i(t); i = 1, \ldots, k; t = 1, \ldots, t\}$ taking values in $\mathbb{R}^k$ with $k$ variables at $n$ time points. We will show in Theorem 3.1 that the arc set of this network describes exactly the conditional dependencies between variables observed at successive time points (e.g., $t - 1$ and $t$) given all other variables observed at the earliest time point (e.g., $t - 1$). This result rests on the three assumptions below.

**Assumption 1** *The stochastic process* $\mathbf{X}$ *is first-order Markovian.*

**Assumption 2** *For all* $t > 0$, *the random variables* $X(t) = (X_1(t), \ldots, X_i(t), \ldots, X_k(t))$ *observed at time* $t$ *are conditionally independent given the random variables* $X(t-1)$ *at the previous time* $t - 1$.

**Assumption 3** *The temporal profile* $(X_i(1), \ldots, X_i(n))$ *of any variable* $X_i$ *cannot be written as a linear combination of the other profiles* $(X_j(1), \ldots, X_j(n))$, $j \neq i$.

Assumption 1 guarantees that any variable at time $t$ is dependent on the past variables only through the variables observed at time $(t - 1)$. On the other hand, Assumption 2 guarantees the variables observed simultaneously at any time point to be conditionally independent given their immediate past. In other words, time points are assumed to be close enough that a variable $X_i$ at time $t$ is better explained by $X(t-1)$ than by other variables $X_j$ at the same time $t$. Therefore, Assumptions 1 and 2 allow the existence of a dynamic Bayesian network with graph $G$ that only contains arcs pointing out from a variable observed at time $(t - 1)$ toward a variable observed at time $t$, with no arcs between simultaneously observed variables. In order to restrict the number of parameters of the network, we assume a constant time delay for all interactions, called the *time point sampling*, defined by the interval between successive time points. It is certainly possible to add simultaneous interactions, or

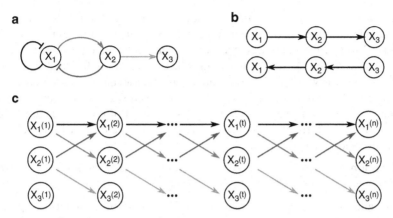

**Fig. 3.2** Graphical representation of a static Bayesian network, a dynamic Bayesian network, and a time-varying dynamic Bayesian network. (a) An example of an interaction network between three variables $X_1$, $X_2$, $X_3$ modeling a genetic regulatory motif between genes. Arcs ending with an arrow correspond to gene activations, and arcs ending with a line correspond to gene inhibitions. (b) Because Bayesian networks are constrained to be acyclic, they cannot contain loops or cycles. Therefore, the motif in (a) cannot be correctly modeled by a conventional (static) Bayesian network which does not take temporal ordering into account. If $X_3$ is conditionally independent from $X_1$ given $X_2$, we have that both $P(X_1, X_2, X_3) = P(X_3 | X_2) P(X_2 | X_1) P(X_1)$ and $P(X_1, X_2, X_3) = P(X_3 | X_2) P(X_1 | X_2) P(X_2)$ are valid sets of local distributions. (c) If time-course expression measurements are available, it is possible to unravel the feedback cycles and the loops over time points. Such time-homogeneous dynamic Bayesian network assumes that at each given time $t$, all the parents of each node are measured at the previous time point $t-1$

a longer time delay, by allowing the existence of arcs between variables observed either at the same time $t$ or with a longer time delay (i.e., from $t-2$ to $t$). However, the number of parameters of the model increases exponentially with the number of time delays, which can be challenging given the number of time points in most data sets.

Finally, Assumption 3 guarantees the uniqueness of $G$ when the $k$ variables are linearly independent, i.e., none of the profiles can be written as a linear combination of the others. When these three assumptions are satisfied, the probability distribution of the process $\mathbf{X}$ can be represented by a dynamic Bayesian network as described by the following theorem.

**Theorem 3.1.** *Whenever Assumptions 1, 2, and 3 are satisfied, the probability distribution of $\mathbf{X}$ can be represented as a dynamic Bayesian network with a directed acyclic graph $G$ whose arcs describe exactly the conditional dependencies between any pair of variables $(X_j(t-1), X_i(t))$ at successive time points given the past variables $X(t-1) \setminus \{X_j(t-1)\}$.*

As expected, dynamic Bayesian network models are dependent on the sampling time and the choice of the time delay. Interactions that occur at a time scale shorter than the sampling time may not necessarily be detected from the given data and can lead to spurious conclusions on the network structure. Thus, a prudent choice of the sampling time may be critical for meaningful results. It might be interesting, for example, to infer a network with a time delay from $t - 1$ to $t$ and another one from $t - 2$ to $t$. If the time delay between two successive time points is too large, considering a static Bayesian network might be a better choice for the data.

In order to carry out model estimation, it is often assumed that the process is homogeneous over time (Assumption 4). In other words, we assume that the phenomenon we are modeling is governed by the same set of rules during the whole experiment. Therefore, $(n - 1)$ repeated measurements are observed for each variable at two successive time points.

**Assumption 4** *The process is homogeneous over time: all arcs in the network and their directions are invariant over time.*

This allows a good representation of a MTS with a limited number of parameters. Each additional time delay would require a specific $k \times k$ coefficient matrix; therefore, a large number of repeated measurements for each variable at a given time point would be needed for estimation. However, such data is rarely available. For instance, most gene expression time series contain no or very few repeated measurements for each gene at a given time point.

While time homogeneity is a strong assumption and not always satisfied for real-world data, it is often used as a simplifying assumption when the number of observations is small compared to the number of variables. For completeness, we also discuss in Sect. 3.4 a recent approach for learning nonhomogeneous dynamic Bayesian network inference that does not impose homogeneity assumptions.

### 3.2.2 Dynamic Bayesian Network Representation of a VAR Process

In dynamic Bayesian networks, it is commonly assumed that dependence relationships are represented by a vector auto-regressive process as defined in Eq. 3.5. A similar assumption characterized static Gaussian Bayesian networks in Sect. 2.2.4. If we assume a VAR process of order 1,

$$X(t) = AX(t - 1) + B + \varepsilon(t) \quad \text{with} \quad \varepsilon(t) \sim N(0, \Sigma), \tag{3.13}$$

then all the arcs are defined between two successive time points. The arc set is defined by the set of nonzero coefficients in $A$; if the element $a_{ij}, i \neq j$ is different from zero, then the network includes an arc from $X_i(t - 1)$ to $X_j(t)$. Furthermore, we assume that the error term for each variable $X_i$ is independent from both the other variables and the respective error terms, so that off-diagonal elements in $\Sigma$ can be set to 0. Of interest is to note that for a VAR(1) process, Assumption 4 is automatically satisfied. The $k \times k$ coefficient matrix $A = (a_{ij})$—which has the same nonzero elements as the adjacency matrix of the interaction network from Fig. 3.2a—and the $k \times 1$ column vector $B = (b_i)$—representing the baseline measurement for each variable—are invariant as a function of time. Moreover, Assumption 1 is satisfied since the random vector $X(t)$ depends only on the random vector at time $(t - 1)$.

Assumption 2 is also satisfied by construction provided the error covariance matrix $\Sigma$ is diagonal (see Lèbre 2009). Assuming uncorrelated errors between different variables may not necessarily hold in real-world scenarios. Nevertheless, it is not unreasonable. Assumption 3 is difficult to verify, but it is not too restrictive if the variables included in the data set are distinct. Then from Theorem 3.1, a VAR(1) process whose error covariance matrix $\Sigma$ is diagonal can be represented by a dynamic Bayesian network whose arcs are identified by the nonzero elements of $A$.

For an illustration, any VAR(1) process with diagonal $\Sigma$ where matrix $A$ has the following form (where the elements $a_{ij}$ refer to nonzero coefficients),

$$A = \begin{pmatrix} a_{11} & a_{12} & 0 \\ a_{21} & 0 & 0 \\ 0 & a_{32} & 0 \end{pmatrix}, \tag{3.14}$$

can be represented by the dynamic network in Fig. 3.2c. For instance, the non-zero coefficient $a_{12}$ implies the arc from $X_2$ to $X_1$ in Fig. 3.2a.

## 3.3 Dynamic Bayesian Network Learning Algorithms

Several approaches have been covered in Chap. 2 for static Bayesian networks. Learning a dynamic Bayesian network defining a VAR model from the given data is a very different process and amounts to identifying the nonzero coefficients of the auto-regressive matrix $A$. Under the homogeneity assumption (Assumption 4 in Sect. 3.2.1), repeated time measurements can be used to perform linear regression. Let $k$ be the number of variables under study. Then each variable $X_i$, $i = 1, \ldots, k$ in a VAR(1) process satisfies

$$X_i(t) = \sum_{j=1}^{k} a_{ij} X_j(t-1) + b_i + \varepsilon_i(t) \quad \text{where} \quad \varepsilon_i(t) \sim N(0, \sigma_i(t)). \tag{3.15}$$

However, the classic ordinary least square estimates of the regression coefficients $a_{ij}$ and $b_i$ can be computed only when $n \gg k$, thus ensuring that the sample covariance matrix has full rank. For real-world data, regularized estimators are required in most cases.

### 3.3.1 Least Absolute Shrinkage and Selection Operator

The *Least Absolute Shrinkage and Selection Operator* or LASSO (Tibshirani 1996) is a standard procedure, first applied to network inference by Meinshausen and Bühlman (2006). This constrained estimation procedure tends to produce some coefficients that are exactly zero by applying an $L_1$ norm penalty to their sum. Variable selection is then straightforward: only nonzero coefficients define significant dependence relationships. Selecting which arcs to include in the network can be done via cross-validation or by minimizing the fraction of the final $L_1$ norm or the mean square error.

### 3.3.2 James–Stein Shrinkage

An efficient estimator of the covariance matrix can be obtained by "shrinking" the empirical correlations coefficients towards zero and the empirical variances to their median. The shrinkage coefficient can be computed in closed form using the expression provided in Ledoit and Wolf (2003), making this approach extremely fast. The resulting correlation matrix has been shown to dominate the empirical one, following classic results on shrinkage from Stein (1956) and James and Stein (1961). Its mean square error is never worse than the mean square error of the empirical correlation matrix.

An application of the James–Stein shrinkage approach to VAR process has been proposed by Opgen-Rhein and Strimmer (2007) and shown to outperform many classic approaches. This can be attributed to improved estimates of the regression coefficients, which are essentially a function of the covariance matrix of $\mathbf{X}$. The network structure is then determined by including the arcs in order of decreasing coefficients. Multiple testing correction to control for false discovery rate (FDR) can also be used with the local FDR approach introduced by Shäfer and Strimmer (2005).

In the context of static Bayesian networks, James–Stein shrinking is used to compute regularized partial correlations and conditional probabilities to use in conditional independence tests. Several such tests are implemented in **bnlearn** for use in constraint-based structure learning algorithms and for independent use via the `ci.test` function.

### 3.3.3 First-Order Conditional Dependencies Approximation

Another powerful approach to learn dynamic Bayesian networks called G1DBN and proposed by Lèbre (2009) is based on first-order conditional dependencies.

The cornerstone of this approach is the concept of *low-order conditional dependence graph*, which originated in the context of the theory of graphical modeling with directed acyclic graphs. The directed acyclic graph defining the dynamic Bayesian network is approximated by the first-order conditional dependencies. Under acceptable conditions, the first-order conditional dependencies graph contains the directed acyclic graph defining the dynamic Bayesian network to be inferred.

By using this approximation, G1DBN implements dynamic Bayesian network learning as a two-step procedure. First, it learns a directed acyclic graph encoding first-order partial dependence relationships. Subsequently, it infers the real network structure of the dynamic Bayesian network using the graph from the previous step. The former is a subgraph of the latter for linear models.

### *3.3.4 Modular Networks*

The *Statistical Inference for MOdular NEtworks* (SIMoNe) by Chiquet et al. (2009) implements various learning algorithms based on the LASSO, with an additional grouping effect for multiple data.

SIMoNe follows a score-based approach; it searches for a latent clustering of the network to drive the selection of arcs through an adaptive $L_1$ penalization of the model likelihood, in particular for VAR(1) processes. The penalization of individual arcs may be weighted according to a predefined latent clustering of the network, thus adapting the inference of the network to a particular topology. The optimal $L_1$ penalty level can be chosen by minimizing the BIC criterion.

Note that this procedure can deal with samples collected in different experimental conditions and therefore not identically distributed.

## 3.4 Non-homogeneous Dynamic Bayesian Network Learning

Homogeneity (Assumption 4 in Sect. 2) is a strong assumption which may not be satisfied for real-world data. For example, the coordination of molecular and bio-chemical processes inside a cell requires highly dynamic gene regulation networks. Different interactions between cellular components can occur across time depending, for instance, on the developmental program of the cell or on physiological and environmental changes. In order to model such data, the ARTIVA approach from Lèbre et al. (2010) uses an *Auto-Regressive TIme VArying* model and the associated learning and inference procedures.

ARTIVA performs an analysis of time-course measurements to identify potential interactions between two sets of variables referred to as *targets* and *parents*, respectively. When considering gene regulation networks, parents are variables whose functions agree with regulatory controls of other genes. Typically these are genes coding for transcription factors. In ARTIVA, each target variable is analyzed independently while searching for dependencies with the parents. As a result, ARTIVA identifies temporal segments in the time-course measurements (Fig. 3.3, dashed lines) for which different interaction models occur between parent and target variables. The time points that delimit the different temporal segments are referred to as *changepoints* (CPs) and define homogeneous *phases*, i.e., sets of time points for which the local network topology (interactions between parent and target genes) remains unchanged.

ARTIVA's probabilistic model is a particular case of a VAR process. Within a temporal phase $h$, each random variable $X_i(t)$ is assumed to depend on its $p$ putative parents, through an auto-regressive model which takes into account a time delay $d$ between the expression values of parent and target genes. The model is defined as

$$X(t) = A^h X(t-1) + B^h + \varepsilon^h(t), \tag{3.16}$$

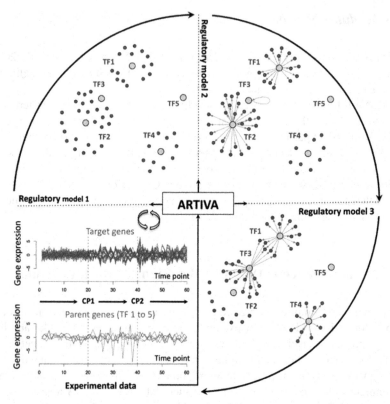

**Fig. 3.3** Time-varying structures of some gene regulation networks learned with ARTIVA. 50 targets (*dark grey*) and 5 transcription factors (parents, *light grey*) are modeled in each network. From an exploratory analysis of the gene expression profiles of the parents and the targets (*bottom left panel*), three temporal phases are identified and delimited by the changepoints CP1 and CP2 (*dashed lines*). The regulatory models for the different temporal phases are represented clockwise in the *top left*, *top right*, and *bottom right* panels. The plots of the networks around each parent were obtained using the IGRAPH library (Csardi and Nepusz 2006). Note that a variable can be simultaneously parent and target, thus allowing the identification of auto-regulation mechanisms as shown here for TF3 (regulatory model 2)

where

- $\varepsilon(t)$ represents the experimental noise and is assumed to follow a $p$-dimensional Gaussian distribution with zero mean and covariance matrix $\Sigma = diag\left((\sigma_i^h)^2\right)$, $\varepsilon_i(t) \sim N(0, (\sigma_i^h)^2)$;
- $B^h$ represents the baseline value of the variables $X_1(t), \ldots, X_p(t)$ in phase $h$ and does not depend on the parents activities;
- $A^h = (a_{ij}^h)$ is a $p \times p$ matrix such that the coefficient $a_{ij}^h$ represents the regulatory interaction between the target $X_i$ and the parent $X_j$ in phase $h$.

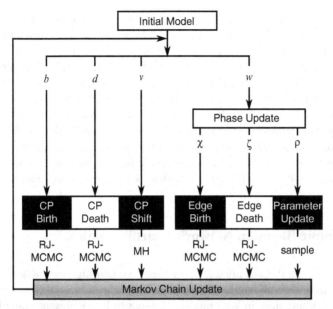

**Fig. 3.4** The ARTIVA algorithm for learning Auto-Regressive TIme VArying networks. The birth, the death, and the shift of a changepoint (CP) are proposed with probabilities $b$, $d$, and $v$, respectively. Updating the regression model describing interactions for a gene within a phase is proposed with probability $w$. Varying the number of CPs or the number of arcs changes the dimension of the state-space and requires RJ-MCMC. Proposed shifts in changepoint positions are accepted according to a standard Metropolis–Hastings step. The probabilities of choosing each modification satisfy $b + d + v + w = 1$ and $\chi + \zeta + \rho = 1$

Each nonzero value in $A^h$ indicates a relationship between the expression levels of two variables $X_i$ and $X_j$ and is therefore a good indicator of a putative biological interaction between those variables. Each of these interactions will be represented with an arc going from a parent $X_j$ at time $t-1$ to a target $X_i$ at time point $t$, for all $t$ in phase $h$. Note that the regulation coefficients are specific to each temporal phase. Finally, for each target $X_i$, the vector of CPs delimiting homogeneous phases is denoted by $\xi^i = (\xi_0^i, \ldots, \xi_{k+1}^i)$, where $\xi_0^i = 1 + d$ and $\xi_{k+1}^i = n + 1$.

In order to learn the auto-regressive time-varying network models, ARTIVA uses the Reversible Jump Markov chain Monte Carlo (RJ-MCMC) approach from Green (1995). RJ-MCMC starts with a randomly generated initial model. At each iteration of the algorithm, a modification of the model is proposed (Fig. 3.4) and can be accepted or rejected with a specific probability, which is computed from the data. The resulting reversible Markov chain sampler can jump between parameter spaces of different dimensions and converges to its stationary distribution after a sufficiently large number of burn-in iterations. After the burn-in, RJ-MCMC provides a good approximation of the probability of each time-varying network model.

The most interesting characteristic of ARTIVA is that it considers *simultaneously* all possible combinations of CPs and all possible network topologies within the different phases. Furthermore, ARTIVA allows network structures associated with different nodes to change with time in different ways, and at the same time it allows each variable to be simultaneously parent and target. This flexibility proves problematic for time series with a low number of measurements, such as the ones typically available in systems biology, leading to overfitting or inflated inference uncertainty. This limitation can be overcome with the semiflexible model proposed in Dondelinger et al. (2013), which is based on a piecewise homogeneous dynamic Bayesian network regularized by gene-specific intersegment information sharing. This approach is implemented in the R package EDISON (Dondelinger et al. 2012).

## 3.5 Dynamic Bayesian Network Learning with R

In this section, we discuss popular R packages for investigating multivariate time series including several diagnostic tests to assess the goodness of fit and justifying the inherent assumptions in VAR models. Subsequently, we present examples using R packages for dynamic Bayesian network modeling. In this regard, we discuss examples from time-homogeneous (Sects. 3.5.2 and 3.3.2) as well as time-varying multivariate time series (Sect. 3.5.4).

### 3.5.1 Multivariate Time Series Analysis

A useful R package for analyzing time series as a VAR processes is **vars**. Consider an example below using the data set Canada with 4 macroeconomic indicators (prod, labor productivity; e, employment; U, unemployment rate; and rw, real wages).

```
> library(vars)
> data(Canada)
```

A vector auto-regressive process VAR($p$) can be fitted from these data with the VAR function.

```
> VAR(Canada, p = 2)
```

The object of class varest returned by VAR contains information on several aspects of the VAR process and its parameters. A summary of the results of VAR can be obtained using the function summary as follows.

```
> summary(VAR(Canada, p = 2))
```

Function VAR has an argument called type which allows the use of different types of deterministic regressors. It defaults to const, which adds the intercept (i.e., the baseline value) to the model. Other possible values are none for a model without the intercept,

```
> VAR(Canada, p = 2, type = "none")
```

trend for a model with a linear trend term,

```
> VAR(Canada, p = 2, type = "trend")
```

and both to add both the intercept and a linear trend to the model.

```
> VAR(Canada, p = 2, type = "both")
```

The p argument specifies the order of the vector auto-regressive process.

The optimal lag order for the VAR process can be estimated using information criteria such as the Akaike Information Criterion (ic = "AIC", the default) or the Schwarz Criterion (ic = "SC"). An upper bound must be specified with the lag.max argument, i.e., lag.max = 4.

```
> VAR(Canada, lag.max = 4, ic = "AIC")
> VAR(Canada, lag.max = 4, ic = "SC")
```

Several approaches to verify the covariance stationarity of a VAR process are implemented in the stability function; the default one computes the *cumulative sums* of the residuals (OLS-CUSUM) of the process, which can be used for an exploratory analysis.

```
> var.2c = VAR(Canada, p = 2, type = "const")
> stab = stability(var.2c, type = "OLS-CUSUM")
> plot(stab)
```

The Jarque–Bera normality tests for univariate and multivariate series are implemented in the normality.test function and are applied to the residuals of the VAR(p). Skewness and kurtosis tests are computed at the same time.

```
> normality.test(var.2c)
```

By default, normality.test computes the multivariate Jarque–Bera test on the standardized residuals. It is important to note that the results of this test depend on the ordering of the variables. The univariate Jarque–Bera tests for the variables in the VAR process can also be computed by setting multivariate.only to FALSE.

```
> normality.test(var.2c, multivariate.only = FALSE)
```

Other diagnostic tests from Sect. 3.1.2, such as testing for the absence of auto-correlation, heteroscedasticity, or non-normality in the errors, are also available in **vars**. For example, the function serial.test implements the Portmanteau test and the Breusch–Godfrey serial correlation Lagrange multiplier test for the serial correlation.

```
> serial.test(var.2c, lags.pt = 16,
+                              type = "PT.adjusted")

Portmanteau Test (adjusted)
data:  Residuals of VAR object var.2c
Chi-squared = 231.5907, df = 224, p-value = 0.3497
```

If the `type` argument is set to `"PT.adjusted"`, the Portmanteau test is computed; the Breusch–Godfrey test is computed if `type` is set to `"BG"`.

In addition, the univariate and multivariate ARCH-LM tests for heteroscedasticity are implemented in `arch.test`.

```
> arch.test(var.2c)

ARCH (multivariate)

data:   Residuals of VAR object var.2c
Chi-squared = 538.8897, df = 500, p-value = 0.1112
```

As was the case for `normality.test`, by default only the multivariate test is computed. We can set `multivariate.only` to `FALSE` to compute the univariate tests as well.

```
> arch.test(var.2c, multivariate.only = FALSE)
```

For a complete overview of the functionality implemented in **vars**, we refer the reader to Pfaff (2008a,b).


### 3.5.2 LASSO Learning: lars and simone

Several implementations of LASSO are available in R; in the following, we will use the **lars** package developed by Hastie and Efron (2012). Other possible choices are **glmnet** by Friedman et al. (2010) and **penalized** by Goeman (2012).

```
> library(lars)
```

Consider the `arth800` MTS data set from the **GeneNet** package. `arth800` describes the temporal expression of 800 genes of the *Arabidopsis thaliana* during the diurnal cycle. In the following example, we will consider a subset `arth12` of 12 genes.

```
> library(GeneNet)
> data(arth800)
> subset = c(60, 141, 260, 333, 365, 424, 441, 512,
+                 521, 578, 789, 799)
> arth12 = arth800.expr[, subset]
```

Model estimation is performed using the `lars` function for a target variable specified by a vector (`y`) and a set of possible parents specified by a matrix of predictors (`x`). The `arth800` data set is composed of 2 time series of 11 time points each: there are two repeated measurements for each time point. Suppose we want to estimate a VAR(1) process. Consequently, we removed the two repeated measurements for the first time point from `y` and the two repeated measurements for the last time point from `x`; they cannot be used for the LASSO due to the lack of the corresponding time points in x and y, respectively.

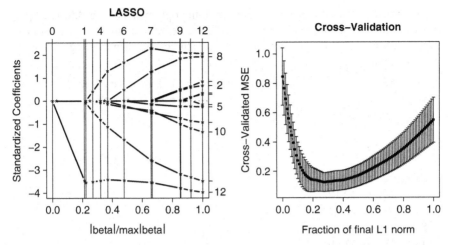

**Fig. 3.5** Graphical output from package **lars**. *Left*: learning process for `lasso.fit`. Arcs are included one at a time, and each inclusion is marked with a vertical line. *Right*: cross-validation estimates of the mean square error as a function of the fraction of the final value of the $L_1$ norm

```
> x = arth12[1:(nrow(arth12) - 2), ]
> y = arth12[-(1:2), "265768_at"]
> lasso.fit = lars(y = y, x = x, type = "lasso")
```

Similarly, we can fit one LASSO model each target variable as follows.

```
> fit.all = lapply(colnames(arth12),
+     function(gene) {
+       y = arth12[-(1:2), gene]
+       lars(y = y, x = x, type = "lasso")
+     })
```

The order in which the coefficients for the parents are included in `lasso.fit`, which is the order in which the corresponding arcs are included in the gene expression network around the gene `265768_at`, is shown in the left panel of Fig. 3.5 and is produced with the `plot` method for `lars` objects.

```
> plot(lasso.fit)
```

In addition, the list of the coefficients fitted at each step of LASSO is given by the `coef` function.

```
> coef(object)
```

Structure learning, which amounts to selecting which arcs should be included in the regulatory network, can be performed via cross-validation with the `cv.lars` function.

```
> lasso.cv = cv.lars(y = y, x = x, mode = "fraction")
```

The graphical output of cv.lars is shown in the right panel of Fig. 3.5. The optimal set of arcs to include in the network is chosen by minimizing the mean square error as a function of the fraction of the final value of the $L_1$ norm (e.g., when all arcs are included).

```
> frac = lasso.cv$index[which.min(lasso.cv$cv)]
> predict(lasso.fit, s = frac, type = "coef",
+    mode = "fraction")
$s
[1] 0.2323232

$fraction
[1] 0.2323232

$mode
[1] "fraction"

$coefficients
  265768_at    263426_at    260676_at    258736_at
-0.04137319  0.00000000   0.00000000   0.00000000
  257710_at    255764_at    255070_at    253425_at
 0.00000000  0.02891478   0.00000000   0.00000000
  253174_at    251324_at    245319_at    245094_at
 0.00000000  0.00000000   0.00000000  -0.72815587
```

The nonzero coefficients in the output of predict indicate which arcs are incident on the gene 265768_at for the optimal fraction s = frac computed by cv.lars.

Structure learning can also be performed by stopping the LASSO estimation after a certain number s of steps (i.e., s = 3) by setting the mode argument of predict to step.

```
> predict(lasso.fit, s = 3, type = "coef",
+    mode = "step")$coefficients
  265768_at    263426_at    260676_at    258736_at
-0.02152962  0.00000000   0.00000000   0.00000000
  257710_at    255764_at    255070_at    253425_at
 0.00000000  0.00000000   0.00000000   0.00000000
  253174_at    251324_at    245319_at    245094_at
 0.00000000  0.00000000   0.00000000  -0.72966658
```

Finally, we can specify the $L_1$ penalty itself, i.e., $s = 0.2$, with mode = "lambda".

```
> predict(lasso.fit, s = 0.2, type = "coef",
+    mode = "lambda")$coefficients
 265768_at  263426_at  260676_at  258736_at
 0.0000000  0.0000000  0.0000000  0.0000000
```

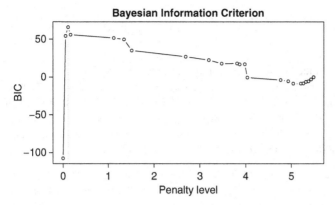

**Fig. 3.6** Graphical output of the **simone** package for dynamic Bayesian network learning via BIC criterion minimization

```
257710_at   255764_at   255070_at   253425_at
0.0000000   0.0000000   0.0000000   0.0000000
253174_at   251324_at   245319_at   245094_at
0.0000000   0.0000000   0.0000000  -0.6961228
```

In addition to the LASSO, the **lars** package implements Least Angle Regression (LAR) from Efron et al. (2004) and stepwise regression. Both can be fitted with the same functions used for the LASSO by setting the `type` arguments of `lars` and `cv.lars` to either `"lar"` or `"stepwise"`, as shown below.

```
> lar.fit = lars(y = y, x = x, type = "lar")
> plot(lar.fit)
> lar.cv = cv.lars(y = y, x = x, type = "lar")
> step.fit = lars(y = y, x = x, type = "stepwise")
> plot(step.fit)
> step.cv = cv.lars(y = y, x = x, type = "stepwise")
```

The **simone** package for Statistical Inference for MOdular NEtworks by Chiquet et al. (2009) provides an implementation of the LASSO specifically targeted to dynamic Bayesian network learning. Model estimation is performed using the `simone` function with `clustering = FALSE`.

```
> library(simone)
> simone(arth12, type = "time-course")
```

The `simone` function allows clustering assumption, i.e., modular network. Model estimation is now performed using the `simone` function with clustering.

```
> ctrl = setOptions(clusters.crit = "BIC")
> simone(arth12, type = "time-course",
+    clustering = TRUE, control = ctrl)
```

The optimal value of the $L_1$ penalty can be chosen by minimizing the BIC criterion, which is computed by `simone` when the model is fitted with `output = "BIC"`.

```
> plot(simone(arth12, type = "time-course",
+           clustering = TRUE, control = ctrl),
+           output = "BIC")
```

The plot produced by the R code above is shown in Fig. 3.6.

Other output options include, among others, output = "AIC" for the AIC criterion and output = "sequence" for stepwise selection. If no output is specified, plot generates a comprehensive set diagnostic plots including a BIC plot, an AIC plot, the regularization paths of the regression coefficients, and the order of inclusion of the arcs.

```
> plot(simone(arth12, type = "time-course",
+           clustering = TRUE, control = ctrl))
```

It is important to note that the **simone** package can learn dynamic Bayesian networks from sets of samples collected under different experimental conditions and therefore not identically distributed. This achieved by adding a grouping effect to the LASSO model and learning multiple related networks in a single call to simone. In that case, approaches such as Group LASSO or Cooperative LASSO are used for learning instead of the original LASSO.

### 3.5.3 Other Shrinkage Approaches: GeneNet, G1DBN

The James–Stein shrinkage estimators proposed by Opgen-Rhein and Strimmer (2007) are implemented in the **GeneNet** package. The VAR coefficients (i.e., the elements of the matrix $A$ in Eq. 3.13) can be robustly estimated by the ggm. estimate.pcor function when the method argument is set to "dynamic".

```
> library(GeneNet)
> dyn = ggm.estimate.pcor(arth, method = "dynamic")
```

Structure learning is carried out by ordering the arcs according to magnitude of their coefficients and performing multiple testing correction with the local FDR approach introduced by Shäfer and Strimmer (2005). Both these tasks can be performed with the network.test.edges function.

```
> arth.arcs = network.test.edges(dyn)
```

We can then identify which arcs are significant with extract.network and include them in the network. Several criteria for the significance threshold are available via the method.ggm argument. For instance, we can just select the top cutoff.ggm arcs with method.ggm = "number".

```
> arth.net = extract.network(arth.edges,
+                   method.ggm = "number", cutoff.ggm = 10)
```

We can also select all the arcs below a chosen threshold (cutoff.ggm) with method.ggm = "prob".

```
> arth.net = extract.network(arth.edges,
+                   method.ggm = "prob", cutoff.ggm = 0.05)
```

Another approach for dynamic Bayesian network learning is implemented in the **G1DBN** package (Lèbre, 2008). To illustrate it, we will use the copy of the arth800 data set included in **G1DBN** under the name arth800line. We choose a subset of this data set to obtain arth12.

```
> library(G1DBN)
> data(arth800line)
> subset = c(60, 141, 260, 333, 365, 424, 441, 512,
+                 521, 578, 789, 799)
> arth12 = as.matrix(arth800line[, subset])
```

Learning is performed in two steps as described in Sect. 3.3.3. First, we learn the graph $G^{(1)}$ encoding the first-order partial dependencies with the DBNScoreStep1 function.

```
> step1 = DBNScoreStep1(arth12, method = "ls")
```

The object returned by DBNScoreStep1 is a list that contains, among other quantities of interest, the matrix of the scores for the first-order dependencies. Part of that matrix is shown below; as can be seen even from these few entries, it is not symmetric in general.

```
> round(step1$S1ls, 2)[1:6, 1:6]
      [,1] [,2] [,3] [,4] [,5] [,6]
[1,] 0.96 0.99 0.95 0.98 0.95 0.95
[2,] 0.60 0.97 0.92 0.79 0.92 0.08
[3,] 0.82 0.91 0.87 0.97 0.83 0.97
[4,] 0.86 0.93 0.98 0.91 0.77 0.44
[5,] 0.99 0.99 0.97 0.69 0.93 0.57
[6,] 0.97 0.97 0.98 0.97 0.92 0.38
```

Starting from the step1 object, we can identify which arcs are significant for a given threshold with the BuildEdges function.

```
> edgesG1 = BuildEdges(score = step1$S1ls,
+                         threshold = 0.50,   prec = 6)
> nrow(edgesG1)
[1] 27
```

In the second step, we learn with DBNScoreStep2 the real structure $G$ of the dynamic Bayesian network, that is, the one encoding the full-order conditional dependencies of the data.

```
> step2 = DBNScoreStep2(step1$S1ls, data = arth12,
+           method = "ls", alpha1 = 0.50)
```

Subsequently, we can identify which arcs are significant in this new network with BuildEdges and a more stringent threshold than the one we used in the first step.

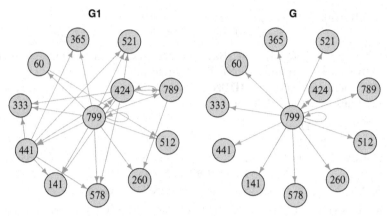

**Fig. 3.7** Network structures learned by **G1DBN** in the first step ($G^{(1)}$, on the *left*) and in the second step ($G$, on the *right*)

```
> edgesG = BuildEdges(score = step2,
+                            threshold = 0.05, prec = 6)
```

The output of the second step is the matrix of the coefficients associated with the arcs of the dynamic Bayesian network; the elements corresponding to arcs not present in the graph are set to NA. The network structures learned in the first and second step of G1DBN are shown in Fig. 3.7. Essentially, the first step (left, Fig. 3.7) performs dimension reduction, while the second step (right, Fig. 3.7) gives the set of arcs defining the network.

### 3.5.4 Non-homogeneous Dynamic Bayesian Network Learning: ARTIVA

The **ARTIVA** package provides several functions for structure learning, parameter learning, and inference in order to facilitate the application of the ARTIVA approach to dynamic Bayesian network learning and the interpretation of the its results.

An example of ARTIVA network learning is given below using a synthetic data set called simulatedProfiles, which contains 55 genes and 30 time points from the **ARTIVA** package.

```
> library(ARTIVA)
> data(simulatedProfiles)
```

Unlike the case of homogeneous dynamic Bayesian network, it is important to identify the target and parent genes prior to learning using simulatedProfiles.

```
> targets = c("1", "10", "20", "TF3", "45", "50")
> parents = c("TF1", "TF2", "TF3", "TF4", "TF5")
```

Then we can call the ARTIVAnet function, specifying the target variables with the targetData argument and parent variables with the parentData argument.

```
> DBN = ARTIVAnet(
+     targetData = simulatedProfiles[targets, ],
+     parentData = simulatedProfiles[parents, ],
+     targetNames = targets,
+     parentNames = parents,
+     niter = 50000,
+     savePictures = FALSE)
```

The number of iterations performed by the algorithm is set with the niter argument.

The return value of ARTIVAnet is a data frame containing, for each pair of parent and target variables and for each phase, the estimated regression coefficient and its posterior probability.

```
> head(ARTIVAtest1[, -7])
  Parent Target CPstart CPend PostProb CoeffMean
1    TF1      1       2    10   0.0469   0.00000
2    TF2      1       2    10   0.0157   0.00000
3    TF3      1       2    10   0.0349   0.00000
4    TF4      1       2    10   0.0317   0.00000
5    TF5      1       2    10   0.0206   0.00000
6    TF1      1      11    30   0.9996  -1.52802
```

When savePictures = FALSE, ARTIVAnet produces several sets of plots detailing the progress of the (RJ-)MCMC simulation. An example is shown in Fig. 3.8. If, on the other hand, savePictures is set to FALSE, the same plots are saved in an output file (by default, a PDF file in an ad hoc subdirectory named ARTIVAnet).

## Exercises

**3.1.** Consider the Canada data set from the **vars** package, which we analyzed in Sect. 3.5.1.

(a) Load the data set from the **vars** package and investigate its properties using the exploratory analysis techniques covered in Chap. 1.
(b) Estimate a VAR(1) process for this data set.
(c) Build the auto-regressive matrix $A$ and the constant matrix $B$ defining the VAR(1) model.
(d) Compare the results with the LASSO matrix when estimating the $L_1$-penalty with cross-validation.
(e) What can you conclude?

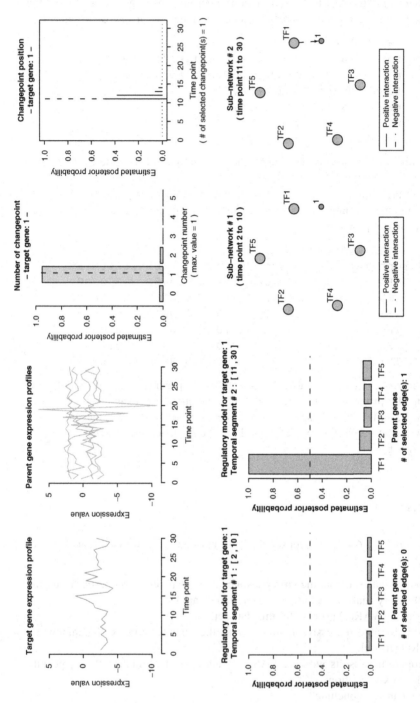

**Fig. 3.8** Graphical outputs obtained from the **ARTIVA** package. Results presented here were obtained analyzing the simulated data sets data (simulatedData) available in the **ARTIVA** package, applying function ARTIVAsubnet. The TF3 was considered simultaneously as parent and target, thus allowing the identification of an auto-regulation mechanism

**3.2.** Consider the `arth800` data set from the **GeneNet** package, which we analyzed in Sects. 3.5.2 and 3.5.3.

(a) Load the data set from the **GeneNet** package. The time series expression of the 800 genes is included in a data set called `arth800.expr`. Investigate its properties using the exploratory analysis techniques covered in Chap. 1.
(b) For this practical exercise, we will work on a subset of variables (one for each gene) having a large variance. Compute the variance of each of the 800 variables, plot the various variance values in decreasing order, and create a data set with the variables greater than 2.
(c) Can you fit a VAR process with usual approach from this data set?
(d) Which alternative approaches can be used to fit a VAR process from this data set?
(e) Estimate a dynamic Bayesian network with each of the alternative approaches presented in this chapter.

**3.3.** Consider the dimension reduction approaches used in the previous exercise and the `arth800` data set from the **GeneNet** package.

(a) For a comparative analysis of the different approaches, select the top 50 edges for each approach (function `BuildEdges` from the **G1DBN** package can be used to that end).
(b) Plot the four inferred networks with the function `plot` from package **G1DBN**.
(c) How many edges are common to the four inferred networks?
(d) Are the top 50 edges of each inferred network similar? What can you conclude?

# Chapter 4
# Bayesian Network Inference Algorithms

**Abstract** Chapters 2 and 3 discussed the importance of learning the structure and the parameters of Bayesian networks from observational and interventional data sets. Bayesian inference on the other hand is often a follow-up to Bayesian network learning and deals with inferring the state of a set of variables given the state of others as evidence. Such an approach eliminates the need for additional experiments and is therefore extremely helpful. In this chapter, we will introduce inferential techniques for static and dynamic Bayesian networks and their applications to gene expression profiles.

## 4.1 Reasoning Under Uncertainty

Bayesian networks, like other statistical models, can be used to answer questions about the nature of the data that go beyond the mere description of the observed sample. Techniques used to obtain those answers based on new evidence are known in general as *inference*. For Bayesian networks, the process of answering these questions is also known as *probabilistic reasoning* or *belief updating*, while the questions themselves are called *queries*. Both names were introduced by Pearl (1988) and borrowed from expert systems theory (e.g., you would submit a *query* to an *expert* to get an opinion and *update your beliefs* accordingly) and have completely replaced traditional statistical terminology in recent works such as Koller and Friedman (2009).

### 4.1.1 Probabilistic Reasoning and Evidence

In practice, probabilistic reasoning on Bayesian networks has its roots embedded in Bayesian statistics and focuses on the computation of posterior probabilities or densities. For example, suppose we have learned a Bayesian network $B$ with

R. Nagarajan et al., *Bayesian Networks in R: with Applications in Systems Biology*,
Use R! 48, DOI 10.1007/978-1-4614-6446-4_4,
© Springer Science+Business Media New York 2013

structure $G$ and parameters $\Theta$, under one of the distributional assumptions detailed in Sect. 2.2.4. Subsequently, we want to investigate the effects of a new piece of *evidence* $\mathbf{E}$ on the distribution of $\mathbf{X}$ using the knowledge encoded in $B$, that is, to investigate the posterior distribution $P(\mathbf{X}|\mathbf{E},B) = P(\mathbf{X}|\mathbf{E},G,\Theta)$.

The approaches used for this kind of analysis vary depending on the nature of $\mathbf{E}$ and on the nature of information we are interested in. The two most common kinds of evidence are as follows:

- *Hard evidence*, an instantiation of one or more variables in the network. In other words,

$$\mathbf{E} = \left\{X_{i_1} = e_1, X_{i_2} = e_2, \ldots, X_{i_k} = e_k\right\}, \qquad i_1, \ldots, i_k \in \{1, \ldots n\}, \qquad (4.1)$$

which ranges from the value of a single variable $X_i$ to a complete specification for $\mathbf{X}$. Such an instantiation may come, for instance, from a new (partial or complete) observation recorded after the Bayesian network was learned.

- *Soft evidence*, a new distribution for one or more variables in the network. Since both the network structure and the distributional assumptions are treated as fixed, soft evidence is usually specified as a new set of parameters,

$$\mathbf{E} = \left\{X_{i_1} \sim (\Theta_{X_{i_1}}), X_{i_2} \sim (\Theta_{X_{i_2}}), \ldots, X_{i_k} \sim (\Theta_{X_{i_k}})\right\}. \qquad (4.2)$$

This new distribution may be, for instance, the null distribution in a hypothesis testing problem.

As far as queries are concerned, we will focus on *conditional probability queries* (CPQ) and *maximum a posteriori* (MAP) queries, also known as *most probable explanation* (MPE) queries. Both apply mainly to hard evidence, even though they can be used in combination with soft evidence.

Conditional probability queries are concerned with the distribution of a subset of variables $\mathbf{Q} = \{X_{j_1}, \ldots, X_{j_l}\}$ given some hard evidence $\mathbf{E}$ on another set $X_{i_1}, \ldots, X_{i_k}$ of variables in $\mathbf{X}$. The two sets of variables are usually assumed to be disjoint. In discrete Bayesian networks, this distribution is computed as the posterior probability

$$CPQ(\mathbf{Q}|\mathbf{E},B) = P(\mathbf{Q}|\mathbf{E},G,\Theta) = P(X_{j_1}, \ldots, X_{j_l}|\mathbf{E},G,\Theta), \qquad (4.3)$$

which is the marginal posterior probability distribution of $\mathbf{Q}$, i.e.,

$$P(\mathbf{Q}|\mathbf{E},G,\Theta) = \int P(\mathbf{X}|\mathbf{E},G,\Theta)\,d(\mathbf{X} \setminus \mathbf{Q}). \qquad (4.4)$$

In Gaussian Bayesian networks, likewise,

$$CPQ(\mathbf{Q}|\mathbf{E},B) = f(\mathbf{Q}|\mathbf{E},G,\Theta) = \int f(\mathbf{X}|\mathbf{E},G,\Theta)\,d(\mathbf{X} \setminus \mathbf{Q}). \qquad (4.5)$$

This class of queries has many useful applications due to their versatility. For instance, conditional probability queries can be used to assess the interplay between two sets of experimental design factors for a given trait of interest. While the latter (i.e., the trait) would be considered as the hard evidence $E$, the former would play the role of the set of query variables $\mathbf{Q}$. Another common example would be assessing the odds of an unfavorable outcome $\mathbf{Q}$ for different sets of hard evidence $\mathbf{E}_1$, $\mathbf{E}_2, \ldots, \mathbf{E}_m$.

Maximum a posteriori queries are concerned with finding the configuration $\mathbf{q}^*$ of the variables in $\mathbf{Q}$ that has the highest posterior probability,

$$MAP(\mathbf{Q}\,|\,\mathbf{E}, B) = \mathbf{q}^* = \underset{\mathbf{q}}{\operatorname{argmax}} \, P(\mathbf{Q} = \mathbf{q}\,|\,\mathbf{E}, G, \Theta) \qquad (4.6)$$

or the maximum posterior density

$$MAP(\mathbf{Q}\,|\,\mathbf{E}, B) = \mathbf{q}^* = \underset{\mathbf{q}}{\operatorname{argmax}} \, f(\mathbf{Q} = \mathbf{q}\,|\,\mathbf{E}, G, \Theta) \qquad (4.7)$$

in Gaussian Bayesian networks. Applications of this kind of query fall into two categories: imputing missing data from partially observed hard evidence, where the variables in $\mathbf{Q}$ are not observed and are to be imputed from the ones in $\mathbf{E}$, or comparing $\mathbf{q}^*$ with the observed values for the variables in $\mathbf{Q}$ for completely observed hard evidence.

Both conditional probability queries and maximum a posteriori queries can also be used with soft evidence, albeit with different interpretations. For instance, when $\mathbf{E}$ encodes hard evidence it is not stochastic but an observed value. In this case, $P(\mathbf{Q} = \mathbf{q}\,|\,\mathbf{E}, G, \Theta)$ is not stochastic. However, when $\mathbf{E}$ encodes soft evidence it is still a random variable, and in turn $P(\mathbf{Q} = \mathbf{q}\,|\,\mathbf{E}, G, \Theta)$ is also stochastic. Therefore, the results from the queries described in this section must be evaluated according to the nature of the evidence they are based on.

### 4.1.2 Algorithms for Belief Updating: Exact and Approximate Inference

The estimation of the posterior probabilities and densities shown in the previous section is a fundamental problem in the evaluation of queries. Queries involving very small probabilities or large networks are particularly problematic even with the best algorithms in literature due to computational and probabilistic challenges. In the worst case, their computational complexity is exponential in the number of variables.

Algorithms for belief updating can be characterized either as *exact* or *approximate*. Both build upon the fundamental properties of Bayesian networks introduced in Sect. 2.1 to avoid the curse of dimensionality through the use of *local computations*, that is, by only using local distributions.

---

**Algorithm 4.1** Junction Tree Clustering Algorithm

---

1. **Moralize:** create the moral graph of the Bayesian network $B$ as illustrated in Sect. 2.1.4.
2. **Triangulate:** break every cycle spanning 4 or more nodes into subcycles of exactly 3 nodes by adding arcs to the moral graph, thus obtaining a *triangulated graph*.
3. **Cliques:** identify the *cliques* of the triangulated graph, i.e., maximal subsets of nodes in which each element is adjacent to all the others.
4. **Junction Tree:** create a tree in which each clique is a node, and adjacent cliques are linked by arcs.
5. **Reparameterize:** use the parameters of the local distributions of $B$ to compute the parameter sets of the compound nodes of the junction tree.

---

For instance, the marginalization in Eq. 4.4 can be rewritten as

$$P(\mathbf{Q}\,|\,\mathbf{E},G,\Theta) = \int P(\mathbf{X}\,|\,\mathbf{E},G,\Theta)\,d(\mathbf{X}\setminus\mathbf{Q}) =$$

$$= \int \left[ \prod_{i=1}^{p} P(X_i\,|\,\mathbf{E},\Pi_{X_i},\Theta_{X_i}) \right] d(\mathbf{X}\setminus\mathbf{Q}) = \prod_{i:X_i\in\mathbf{Q}} \int P(X_i\,|\,\mathbf{E},\Pi_{X_i},\Theta_{X_i})\,dX_i. \quad (4.8)$$

The correspondence between d-separation and conditional independence can also be used to further reduce the dimension of the problem. From Definition 2.2, variables that are d-separated from $\mathbf{Q}$ by $\mathbf{E}$ cannot influence the outcome of the query. Therefore, they may be completely disregarded in computing the posterior probabilities.

Exact inference algorithms combine repeated applications of Bayes' theorem with local computations to obtain exact values $P(\mathbf{Q}\,|\,\mathbf{E},G,\Theta)$ or $f(\mathbf{Q}\,|\,\mathbf{E},G,\Theta)$. However, their feasibility is restricted to small or very simple networks such as trees and polytrees.

The two best-known exact inference algorithms are *variable elimination* and belief updates based on *junction trees*. Both were originally derived for discrete networks and have been later extended to the continuous and mixed networks. Variable elimination uses the structure of the Bayesian network directly, specifying the optimal sequence of operations on the local distributions and how to cache intermediate results to avoid unnecessary computations. On the other hand, belief updates can also be performed by transforming the Bayesian network into a junction tree first. As illustrated in Algorithm 4.1, a junction tree is a transformation of the moral graph of $B$ in which the original nodes are clustered to reduce any network structure into a tree. Subsequently, belief updates can be performed efficiently using Kim and Pearl's Message-Passing algorithm. The derivation and the details of the major steps

---

**Algorithm 4.2** Logic Sampling Algorithm

---

1. Order the variables in $\mathbf{X}$ according to the topological ordering implied by $G$, say $X_{(1)} \prec X_{(2)} \prec \ldots \prec X_{(p)}$.
2. For a suitably large number of samples $\mathbf{x}^* = (x_1^*, \ldots, x_p^*)$:

    a. for $i = 1, \ldots, p$, generate $x_{(i)}^*$ from $X_{(i)} \mid \Pi_{X_{(i)}}$;
    b. if $\mathbf{x}$ includes $\mathbf{E}$, set $n_{\mathbf{E}} = n_{\mathbf{E}} + 1$;
    c. if $\mathbf{x}$ includes both $\mathbf{Q} = \mathbf{q}$ and $\mathbf{E}$, set $n_{\mathbf{E},\mathbf{q}} = n_{\mathbf{E},\mathbf{q}} + 1$.

3. Estimate $P(\mathbf{Q} \mid \mathbf{E}, G, \Theta)$ with $n_{\mathbf{E},\mathbf{q}}/n_{\mathbf{E}}$.

---

of this algorithm are beyond the scope of this book. A detailed explanation along with step-by-step examples can be found in Korb and Nicholson (2010) and Koller and Friedman (2009).

Approximate inference algorithms use Monte Carlo simulations to sample from the local distributions and thus estimate $P(\mathbf{Q} \mid \mathbf{E}, G, \Theta)$ or $f(\mathbf{Q} \mid \mathbf{E}, G, \Theta)$. In particular, they generate a large number of samples from $B$ and estimate the relevant conditional probabilities by weighting the samples that include both $\mathbf{E}$ and $\mathbf{Q} = \mathbf{q}$ against those that include only $\mathbf{E}$. In computer science, these random samples are often called *particles*, and the algorithms that make use of them are known as *particle filters* or *particle-based methods*.

Many approaches have been developed for both random sampling, weighting, and their combination. This has resulted in several approximate algorithms. Random sampling ranges from the generation of independent samples to more complex Markov chain Monte Carlo (MCMC) schemes. For a gentle introduction to the subject, we refer the reader to Robert and Casella (2009). Common choices are either *rejection sampling* or *importance sampling*. Furthermore, weight functions range from the uniform distribution to likelihood functions to various estimates of posterior probability. The simplest combination of these sampling and weighting approaches is known as either *forward* or *logic sampling*. It is described in Algorithm 4.2 and illustrated in detail in both Korb and Nicholson (2010) and Koller and Friedman (2009). Logic sampling combines rejection sampling and uniform weights, essentially counting the proportion of generated samples including $\mathbf{E}$ that also include $\mathbf{Q} = \mathbf{q}$. Clearly, such an algorithm can be very inefficient if $P(\mathbf{E})$ is small, because most particles will be discarded without contributing to the estimation of $P(\mathbf{Q} \mid \mathbf{E}, G, \Theta)$. However, its simplicity makes it easy to implement and very general in its application; it allows for very complex specifications of $\mathbf{E}$ and $\mathbf{Q}$ for both $MAP(\mathbf{Q} \mid \mathbf{E}, B)$ and $CPQ(\mathbf{Q} \mid \mathbf{E}, B)$. At the other end of the spectrum, complex approximate algorithms such as the *adaptive importance sampling* scheme by Cheng and Druzdel (2000) can estimate conditional probabilities as small as $10^{-41}$. They also perform better on large networks. However, their assumptions often restrict them to discrete data and may require the specification of nontrivial tuning parameters.

## 4.1.3 Causal Inference

When a Bayesian network is given a causal interpretation, the interpretation of queries and evidence changes as well. Just as the arcs in the network describe causal relationships instead of probabilistic dependencies, queries evaluate the probability of known causes given their effects or vice versa.

In this setting, posterior probabilities are not interpreted in terms of beliefs changing according to some observed evidence but rather as measures of the effects of *interventions* on the causal structure. To distinguish the latter from the former, we will denote interventions with $\mathbf{I}$ while keeping the same general notation for both. Interventions play the same role that evidence had in Sect. 4.1.1, and like evidence, they can be classified either as *ideal (perfect) interventions* or *stochastic (imperfect) interventions* (Korb et al., 2004).

Ideal interventions represent the causal analogous of hard evidence; they describe an action whose only effect is to fix the values of the variables in $\mathbf{I}$ to particular set of values

$$\mathbf{I} = \left\{ X_{i_1} = x_1, X_{i_2} = x_2, \ldots, X_{i_k} = x_k \right\}. \tag{4.9}$$

Conditional probability queries of the form

$$P(\mathbf{Q} \mid \mathbf{I}, G, \Theta) = P(X_{j_1}, \ldots, X_{j_l} \mid \mathbf{I}, G, \Theta), \tag{4.10}$$

involving ideal interventions are called *intervention queries*. They evaluate the consequences of the intervention $\mathbf{I}$ on $\mathbf{Q}$ through its posterior distribution. If some hard evidence on a third set of variables is included in the query as well, so that

$$P(\mathbf{Q} \mid \mathbf{I}, \mathbf{E}, G, \Theta) = P(X_{j_1}, \ldots, X_{j_l} \mid \mathbf{I}, \mathbf{E}, G, \Theta), \tag{4.11}$$

the query is called a *counterfactual query*, and it evaluates the consequences of intervention $\mathbf{I}$ in a particular scenario defined by the hard evidence $\mathbf{E}$. In other words, it evaluates the consequences of $\mathbf{I}$ in an alternate world in which $\mathbf{E}$ happened instead of the values actually observed for the sample; hence the name.

Stochastic interventions are very difficult to handle in their most general form. Unlike soft evidence, not only the variables in $\mathbf{I}$ are not fixed, but the set of variables that are included in $\mathbf{I}$ is a random variable. For this reason, they are rarely used in practice even under the simplifying assumption that the set $\mathbf{I}$ is not random. In most cases, assuming that interventions are ideal results in significant computational savings without noticeably degrading the quality of the query. This is the case, for example, in the protein-signaling data from Sachs et al. (2005) studied in Sect. 2.5. Even though the stimulatory cues and the inhibitory interventions applied to the various parts of the data set are hardly ideal, they are assumed to be so to include their effects in the structure learning process. The conclusions of the original paper indicate how this assumption did not invalidate the results of structure learning, but improved its ability to correctly identify causal relationships instead.

## 4.2  Inference in Static Bayesian Networks

Consider again the protein-signaling network and the data set from Sachs et al. (2005) we analyzed in Sect. 2.5.

```
> isachs = read.table("sachs.interventional.txt",
+                 header = TRUE, colClasses = "factor")
```

Before we can use either **bnlearn** or **gRain** to apply the approaches illustrated in Sect. 4.1, we need to create a bn object for the validated network structure from Sachs et al. (2005) and perform parameter learning.

```
> library(gRain)
> library(bnlearn)
> val.str = paste("[PKC][PKA|PKC][praf|PKC:PKA]",
+                 "[pmek|PKC:PKA:praf][p44.42|pmek:PKA]",
+                 "[pakts473|p44.42:PKA][P38|PKC:PKA]",
+                 "[pjnk|PKC:PKA][plcg][PIP3|plcg]",
+                 "[PIP2|plcg:PIP3]")
> val = model2network(val.str)
> isachs = isachs[, 1:11]
> for (i in names(isachs))
+   levels(isachs[, i]) = c("LOW", "AVG", "HIGH")
> fitted = bn.fit(val, isachs, method = "bayes")
```

The INT variable, which codifies the intervention applied to each observation, is not needed for inference and is therefore dropped from the data set. Furthermore, we rename the expression levels of each protein to make both the subsequent R code and its output more readable.

The reason for setting method to "bayes" in bn.fit is twofold. First, Bayesian estimates for the parameters of the network are smoother than the maximum likelihood ones, making inference both easier and more robust. Furthermore, Koller and Friedman (2009) showed that such estimates produce Bayesian networks that are close to the "true" networks for small imaginary sample sizes. On a related note, using Bayesian parameter estimates also guarantees that conditional probability tables are always completely specified (i.e., without missing values) even for small data sets.

### 4.2.1  Exact Inference

In their paper, Sachs et al. (2005) performed two conditional probability queries using the validated Bayesian network val:

1. A direct perturbation of p44.42 should influence pakts473.
2. A direct perturbation of p44.42 should not influence PKA.

The resulting posterior distributions were then compared with the results of two ad hoc experiments to confirm the validity and the direction of the inferred causal influences.

Given the size of the network, we can perform both queries using any of the exact and approximate inference algorithms introduced in Sect. 4.1.2. First, we will start with the implementation of the junction tree algorithm provided by the **gRain** package.

```
> jtree = compile(as.grain(fitted))
```

The `compile` function performs all the steps shown in Algorithm 4.1 and takes the `grain` object returned by `as.grain` as an argument. The latter function, along with `as.bn.fit`, provides an easy way to export Bayesian networks from **bnlearn** to **gRain** and vice versa, thus integrating the functionality of these two packages.

We can then introduce the direct perturbation of `p44.42` required by both queries by calling `setFinding` as follows (Fig. 4.1). In causal terms, this would be an ideal intervention.

```
> jprop = setFinding(jtree, nodes = "p44.42",
+                    states = "LOW")
```

As we can see from the code below, the marginal distribution of `pakts473` is similar whether or not we take the evidence (intervention) into account.

```
> querygrain(jtree, nodes = "pakts473")$pakts473
pakts473
        LOW        AVG        HIGH
0.60893407 0.31041282 0.08065311
> querygrain(jprop, nodes = "pakts473")$pakts473
pakts473
         LOW        AVG        HIGH
0.665161776 0.333333333 0.001504891
```

The slight inhibition of `packts473` resulting from the inhibition of `p44.42` agrees with both the direction of the arc linking the two nodes and the additional experiments performed by Sachs et al. (2005). In causal terms, the fact that changes in `p44.42` affect `packts473` supports the existence of a causal link from the former to the latter.

As far as `PKA` is concerned, both the validated network and the additional experimental evidence support the existence of a causal link from `PKA` to `p44.42`. Therefore, interventions to `p44.42` cannot affect `PKA`. However, knowledge of the expression level of `p44.42` may still alter our expectations on `PKA` if we treat it as evidence instead of an ideal intervention.

```
> querygrain(jtree, nodes = "PKA")$PKA
PKA
       LOW        AVG       HIGH
0.1943315 0.6956254 0.1100431
```

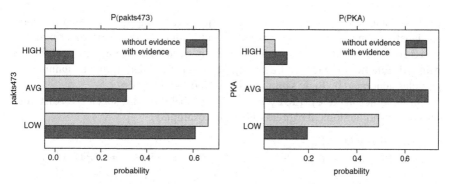

**Fig. 4.1** Probability distributions of pakts473 and PKA before and after inhibiting p44.42

```
> querygrain(jprop, nodes = "PKA")$PKA
PKA
        LOW        AVG       HIGH
0.48908954 0.45116629 0.05974417
```

All the queries illustrated above can be easily changed to maximum a posteriori queries by finding the largest element in the distribution of the target node.

```
> names(which.max(querygrain(jprop,
+                     nodes = c("PKA"))$PKA))
[1] "LOW"
```

## 4.2.2 Approximate Inference

The conditional probability queries from the previous section produce similar results when they are performed with the logic sampling algorithm, illustrated in Algorithm 4.2 and implemented in **bnlearn** in the cpdist and cpquery functions.

```
> particles = cpdist(fitted, nodes = "pakts473",
+                     evidence = (p44.42 == "LOW"))
> prop.table(table(particles))
particles
         LOW         AVG        HIGH
0.665622103 0.332827458 0.001550438
> particles = cpdist(fitted, nodes = "PKA",
+                     evidence = (p44.42 == "LOW"))

> prop.table(table(particles))
particles
        LOW         AVG        HIGH
0.48865276 0.45108553 0.06026171
```

cpdist takes as arguments a bn.fit object describing the Bayesian network, the labels of one or more query nodes, and a logical expression describing the evidence. The latter works in the same way as the analogous argument for the subset function in package base. cpdist returns a data frame containing the particles generated by logic sampling that include evidence.

On the other hand, cpquery returns the probability of a specific event, described by another logical expression. So, for example,

```
> cpquery(fitted,
+    event = (pakts473 == "LOW") & (PKA != "HIGH"),
+    evidence = (p44.42 == "LOW") | (praf == "LOW"))
[1] 0.5594823
```

The combination of events and evidence on different variables through the use of vectorized operators such as !=, ==, &, |, %in% provides a versatile interface for specifying conditional probability queries. This is particularly important when performing inference on Gaussian Bayesian networks, because in this setting both event and evidence are regions in a real space. Therefore, complex combinations of <, <=, >=, and > are required to describe them.

## 4.3 Inference in Dynamic Bayesian Networks

Techniques for learning dynamic Bayesian networks are based on the same fundamental ideas as the ones for learning static networks, as we have seen in Chap. 3 for the dynamic Bayesian networks based on VAR models. The same is true for inference. The most common type of query for such models is to compute the marginal distribution of a node $X_i$ at a time $t$ conditional on other nodes at times $1, \ldots, T$:

- If $T = t$, the query is called *filtering* and consists in querying the state of the network at the current time given all the available information.
- If $T > t$, the query is called *smoothing* and consists in reducing or removing noise from past time points using the information we have collected in the mean time.
- If $T < t$, the query is a *prediction*.

Several exact and approximate inference algorithms specific to dynamic Bayesian networks have been presented in literature. Popular ones include the forward–backward algorithm, the frontier algorithm, the interface algorithm, the Boyen–Koller (BK) algorithm, the Factored Frontier (FF) algorithm, and the Loopy Belief Propagation (LBP) algorithm. For an overview of such approaches, we refer the reader to Murphy's PhD thesis (Murphy 2002). However, the techniques we have been using in the previous section can also be applied to dynamic Bayesian networks.

Most probable explanation queries can be performed for all of filtering, smoothing, and predictions, as shown in Fig. 4.2 for the LASSO model fitted from the arth12 data set with **lars** in Sect. 3.5.2.

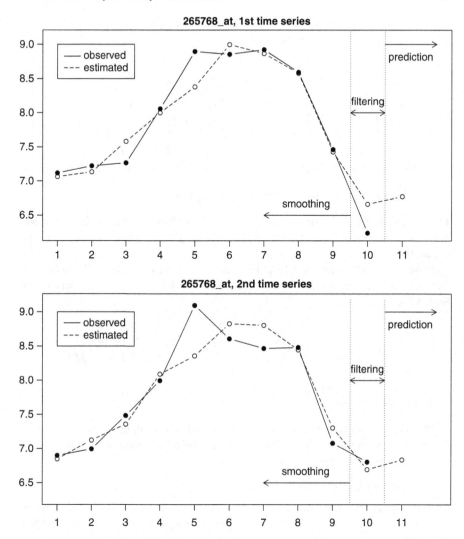

**Fig. 4.2** Observed and estimated expression levels for two time series available for gene 265768_at in the LASSO model from Sect. 3.5.2

```
> x = arth12[1:(nrow(arth12) - 2), ]
> y = arth12[-(1:2), "265768_at"]
> lasso.fit = lars(y = y, x = x, type = "lasso")
```

If we tune the model to find the optimal value for the $L_1$ penalty, we can then estimate the expression levels of the gene 265768_at for all past and present time points.

```
> lasso.cv = cv.lars(y = y, x = x, mode = "fraction")
> frac = lasso.cv$index[which.min(lasso.cv$cv)]
```

```
> lasso.est = predict(lasso.fit, type = "fit",
+                     newx = x, s = frac,
+                     mode = "fraction")$fit
> lasso.est
      0-1         0-2         1-1         1-2         2-1
7.099782    6.894064    7.166249    7.157744    7.592092
      2-2         4-1         4-2         8-1         8-2
7.379432    7.990548    8.078921    8.353137    8.333108
     12-1        12-2        13-1        13-2        14-1
8.940241    8.780302    8.816387    8.758480    8.542374
     14-2        16-1        16-2        20-1        20-2
8.417818    7.446577    7.329513    6.717392    6.747178
```

The expression levels for times 20-1 and 20-2 result from filtering the values for the current time point (e.g., $t$); all the other expression levels are smoothed estimates of past time points (e.g., $1, \ldots, t-1$).

Furthermore, the expression level for gene 265768_at at time $t+1$ can be predicted using the data points we discarded when performing structure learning in Sect. 3.5.2.

```
> lasso.pred = predict(lasso.fit, type = "fit",
+                      newx = arth12[c("24-1", "24-2"), ],
+                      s = frac, mode = "fraction")$fit
> lasso.pred
      24-1        24-2
6.822643    6.882054
```

We can also use cpquery and cpdist to perform complex conditional probability queries; in this case, we will use the **penalized** package to fit the LASSO models because of its integration with **bnlearn**.

```
> library(penalized)
```

Consider again the expression level of gene 265768_at at the current time point $t$. First, we estimate the optimal value of the $L_1$ penalty $\lambda$ and we fit the LASSO model one more time.

```
> lambda = optL1(response = y, penalized = x)$lambda
> lasso.t = penalized(response = y, penalized = x,
+                     lambda1 = lambda)
> coef(lasso.t)
(Intercept)      245094_at
14.0402894    -0.7059011
```

As we can see from the output, the only parent of gene 265768_at is gene 245094_at. The latter seems to inhibit the expression of the former.

Subsequently, we can create the network structure with modelstring and provide the parameters via the custom.fit function. The parameter sets of the nodes are specified by the dist argument, which is a list with one element for each node.

```
> dbn1 =
+    model2network("[245094_at][265768_at|245094_at]")
> xp.mean = mean(x[, "245094_at"])
> xp.sd = sd(x[, "245094_at"])
> dbn1.fit =
+    custom.fit(dbn1,
+        dist = list("245094_at" = list(coef = xp.mean,
+                    sd = xp.sd), "265768_at" = lasso.t))
```

Since we are modeling continuous data, we create a Gaussian Bayesian network. For
the distribution of gene 245094_at, we only need to specify the mean (xp.mean)
and the standard deviation (xp.sd) since the corresponding node has no parents
in dbn1. For gene 265768_at, we reuse the parameters we estimated for the
lasso.t model.

As expected from the regression coefficient in lasso.t, high expression levels
of gene 245094_at at time $t - 1$ make high expression levels of gene 265768_at
at time $t$ much less likely than lower expression levels of gene 245094_at at time
$t - 1$.

```
> cpquery(dbn1.fit, event = ('265768_at' > 8),
+                   evidence = ('245094_at' > 8))
[1] 0.2448749
> cpquery(dbn1.fit, event = ('265768_at' > 8),
+                   evidence = ('245094_at' < 8))
[1] 0.9827778
```

It is important to note that both event and evidence must specify interval events,
because any single point in $\mathbb{R}$ has probability zero. Therefore, any conditional probability query based on such an event will always return to zero.

For a graphical comparison, we can use cpdist to generate two sets of random
observations under the different conditioning events and compare their densities.
The resulting plot is shown in Fig. 4.3.

```
> dist.low = cpdist(dbn1.fit, node = "265768_at",
+              evidence = ('245094_at' < 8))
> dist.high = cpdist(dbn1.fit, node = "265768_at",
+              evidence = ('245094_at' > 8))
```

Performing conditional probability queries using time points that are farther
apart, i.e., $s = t - 2$ and $t$, is also possible. Consider one more time the expression
of gene 265768_at at time $t$; the variables at time $t - 2$ that are most relevant for
our queries are the parents of gene 245094_at as identified by the LASSO model
having 245094_at as a response variable. To avoid name clashes, the expression
level of gene 245094_at at time $t - 2$ will be called 245094_at1.

```
> y = arth12[-(1:2), "245094_at"]
> colnames(x)[12] = "245094_at1"
> lambda = optL1(response = y, penalized = x)$lambda
```

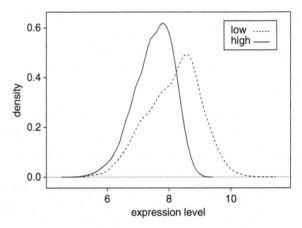

**Fig. 4.3** Density plot of the expression levels of gene 265768_at at time *t* for expression levels of gene 245094_at above 8 (*solid line*) and below 8 (*dashed line*) at time *t* − 1

```
> lasso.s = penalized(response = y, penalized = x,
+                 lambda1 = lambda)
> coef(lasso.s)
  (Intercept)         258736_at       257710_at       255070_at
-2.659077706 -0.009220815   0.273648262 -0.444106451
     245319_at       245094_at1
-0.134050990   1.589716443
```

Here we are assuming that the dynamic Bayesian network is time-homogeneous, since we are using the same data to fit both the variables at time *t* against the ones at time *t* − 1 and the variables at time *t* − 1 against the ones at time *t* − 2.

Subsequently, we create the network structure for the dynamic Bayesian network as we did in the previous example; the result is shown in Fig. 4.4.

```
> dbn2 = empty.graph(c("265768_at", "245094_at",
+         "258736_at", "257710_at", "255070_at",
+         "245319_at", "245094_at1"))
> dbn2 = set.arc(dbn2, "245094_at", "265768_at")
> for (node in names(coef(lasso.s))[-c(1, 6)])
+   dbn2 = set.arc(dbn2, node, "245094_at")
> dbn2 = set.arc(dbn2, "245094_at1", "245094_at")
```

The easiest way to fit the parameters of dbn2 is to estimate all of them via maximum likelihood and then to substitute the parameters of 265768_at and 245094_at with the ones from the LASSO models lasso.t and lasso.s.

```
> dbn2.data = as.data.frame(x[, nodes(dbn2)[1:6]])
> dbn2.data[, "245094_at"] = y
> dbn2.data[, "245094_at1"] = x[, "245094_at"]
> dbn2.fit = bn.fit(dbn2, dbn2.data)
```

```
> dbn2.fit[["265768_at"]] = lasso.t
> dbn2.fit[["245094_at"]] = lasso.s
```

Using the fitted network dbn2.fit, we can now call both cpquery and cpdist to perform smoothing, filtering, and prediction. We may be interested, for example, in the inhibitory effects of prolonged high levels of expression of gene 245094_at (at times $t-2$ and $t-1$) on gene 265768_at.

```
> cpquery(dbn2.fit, event = ('265768_at' > 8),
+   evidence = ('245094_at' > 8) & ('245094_at1' > 8))
[1] 0.1554545
```

This probability is much lower than the corresponding probability computed conditioning only over time $t-1$ (0.2448749), supporting the hypothesis that the inhibitory effects of 245094_at are protracted over time.

This hypothesis is also supported by the fact that, regardless of the expression levels of gene 245094_at at time $t-1$, conditioning on high expression levels at time $t-2$ results in a much lower probability of high expression of gene 265768_at at time $t$ (compared to the unconditional probability, computed by setting evidence = TRUE).

```
> cpquery(dbn2.fit, event = ('265768_at' > 8),
+   evidence = ('245094_at1' > 7) & ('245094_at1' < 8))
[1] 0.9555499
> cpquery(dbn2.fit, event = ('265768_at' > 8),
+   evidence = TRUE)
[1] 0.4507846
```

Finally, given our knowledge on gene 265768_at and gene 245094_at at times $t-2$ and $t$, we can investigate the distribution of 245094_at at time $t-1$.

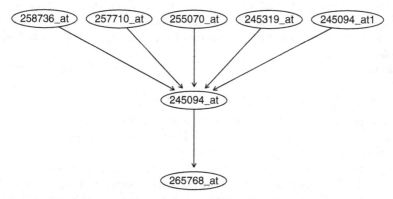

**Fig. 4.4** Dynamic Bayesian network for the expression levels of gene 265768_at going back from time $t$ to time $t-2$

```
> cpd =
+   cpdist(dbn2.fit, node = "245094_at", evidence =
+       ('245094_at1' > 6.5) & ('245094_at1' < 7.5) &
+       ('265768_at' > 7) & ('265768_at' < 8))
> summary(cpd)
   245094_at
 Min.   :7.874
 1st Qu.:8.209
 Median :8.419
 Mean   :8.428
 3rd Qu.:8.612
 Max.   :9.046
```

If the expression level of 245094_at at time $t − 1$ has been observed, we are
smoothing it; if it were missing, we would be imputing it with the mean of its conditional distribution (8.428), that is, its most probable explanation.

## Exercises

**4.1.** Apply the junction tree algorithm to the validated network structure from Sachs
et al. (2005), and draw the resulting undirected triangulated graph.

**4.2.** Consider the Sachs et al. (2005) data used in Sect. 4.2.

(a) Perform parameter learning with the bn.fit function from **bnlearn** and the
    validated network structure. How do the maximum likelihood estimates differ
    from the Bayesian ones, and how do the latter vary as the imaginary sample size
    increases?
(b) Node PKA is parent of all the nodes in the praf → pmek → p44.42 →
    pakts473 chain. Use the junction tree algorithm to explore how our beliefs
    on those nodes change when we have evidence that PKA is "LOW," and when
    PKA is "HIGH."
(c) Similarly, explore the effects on pjnk of evidence on PIP2, PIP3, and plcg.

**4.3.** Consider the marks data set analyzed in Sect. 2.3.

(a) Learn both the network structure and the parameters with likelihood-based ap-
    proaches, i.e., BIC or AIC, for structure learning and maximum likelihood esti-
    mates for the parameters.
(b) Query the network learned in the previous point for the probability to have the
    marks for both STAT and MECH above 60, given evidence that the mark for ALG
    is at most 60. Are the two variables independent given the evidence on ALG?

(c) What is the (conditional) probability of having an average vote (in the $[60, 70]$ range) in both VECT and MECH while having an outstanding vote in ALG (at least 90)?

**4.4.** Using the dynamic Bayesian network dbn2 from Sect. 4.3, investigate the effects of genes 257710_at and 255070_at observed at time $t - 2$ on gene 265768_at at time $t$.

# Chapter 5
# Parallel Computing for Bayesian Networks

**Abstract** Most problems in Bayesian network theory have a computational complexity that, in the worst case, scales exponentially with the number of variables. It is polynomial even for sparse networks. Even though newer algorithms are designed to improve scalability, it is unfeasible to analyze data containing more than a few hundreds of variables. Parallel computing provides a way to address this problem by making better use of modern hardware.

In this chapter we will provide a brief overview of the history and the fundamental concepts of parallel computing, and we will examine their applications to Bayesian network learning and inference using the **bnlearn** package.

## 5.1 Foundations of Parallel Computing

A simple yet effective way to evaluate the performance of a computer program implementing an algorithm is to measure its execution time and the resources it requires to execute successfully. In this respect, performance is influenced by several factors, both hardware and software.

The hardware a program runs on obviously has a profound influence on the program's execution time. Its performance is usually referred to as the *raw performance of the hardware* and is measured by the number of operations it can execute in a given amount of time. Two measures of this kind are the *Input/Output Operations Per Second* (IOPS) for the speed of nonvolatile storage and the *Floating Point Operations Per Second* (FLOPS) for the speed of operations on real numbers. Raw performance is limited by the constraints imposed by the hardware production process, such as the resolution of the lithography techniques used for printing processors on silicon, and increasingly by fundamental physical laws, such as the speed of light and the physics of heat dissipation.

The performance of the software implementation of an algorithm depends both on the *computational complexity* of the algorithm and on the *software architecture* of the implementation itself. Computational complexity classifies algorithms

R. Nagarajan et al., *Bayesian Networks in R: with Applications in Systems Biology*,
Use R! 48, DOI 10.1007/978-1-4614-6446-4_5,
© Springer Science+Business Media New York 2013

according to their inherent difficulty, with particular attention to their behavior as the size of the input grows (*scalability*). Therefore, the only way to improve scalability is to develop a better algorithm for the problem at hand. This is often not possible, either because we are already using an optimal algorithm or because the problem at hand cannot be solved in an *efficient* way (i.e., its computational complexity is more than polynomial in the size of the input in the worst case). Such problems are known as *NP-hard* and are common in the theory of Bayesian networks, both for structure learning (Chickering 1996) and inference (Cooper 1990).

On the other hand, the software architecture can make a significant difference in the overall performance of the program. For instance, choosing the right data structures for the problem can significantly improve or degrade the computational complexity of an algorithm. Tailoring the implementation to the specific hardware it will run on can also result in noticeable speedups.

Parallel computing, defined as the execution of several calculations simultaneously, is an application of this last idea. It was originally introduced to overcome hardware limitations in terms of computational power; large problems were divided into smaller ones, which were then solved concurrently on multipro-cessor supercomputers. Special-purpose hardware architectures were then devel-oped to take advantage of such software, thus maximizing the impact of the raw performance of the hardware for properly implemented programs. A classification of these hardware architectures was proposed by Flynn (1972), according to the nature of the data and the operations they support:

- *Single-Instruction, Single-Data* (SISD): a single processing unit performing a single operations on the same data
- *Multiple-Instruction, Single-Data* (MISD): multiple processing units performing different operations (independently and asynchronously) on the same data
- *Single-Instruction, Multiple-Data* (SIMD): multiple processing units performing the same operation on multiple data
- *Multiple-Instruction, Multiple-Data* (MIMD): multiple processing units perform-ing different operations on multiple data

Nearly all general-purpose modern computers are based on the MIMD model; a notable exception are those that offload part of their workload to *graphical process-ing units* (GPU), as the latter are based on a SIMD model. All modern processors have more than one core, and each core supports multiple concurrent execution threads. Furthermore, in the last few years the speed of processors has peaked be-cause some of the physical constraints mentioned above are preventing further fre-quency scaling. At the same time, the number of cores present in each processor is still increasing, and the same holds for the number of threads supported by each core. As a result, modern computers are remarkably similar to the MIMD parallel architectures from the 1970s and the 1980s described by Flynn, albeit with vastly different capabilities.

This evolution has sparked a renewed interest in parallel computing. Recent re-search efforts focused on two major areas: the *parallelization* of existing algorithms and the development of new algorithms and software libraries explicitly designed to take advantage of parallel computing. However, it is important to note that the degree to which an algorithm can leverage parallel processing depends on the nature of the problem it is trying to address. Some problems are *embarrassingly parallel*, that is, they can be split in such a way that each part rarely or never has to communicate with the other parts. Other problems cannot be fully parallelized, because their parts have to communicate periodically with each other to synchronize their state. If frequent synchronizations are required we speak of *fine-grained parallelism*, and of *coarse-grained parallelism* if synchronizations are only needed a few times over a long period of time. Finally, some problems are *inherently sequential* and cannot be parallelized at all.

## 5.2 Parallel Programming in R

The R interpreter can only execute one command at a time. The only functions that can take advantage of multiple processors are the linear algebra routines provided by the *Basic Linear Algebra Subprograms* (BLAS) library. To this end, R must be compiled against a third-party, multi-threaded implementation of the BLAS library such as the one provided by Intel. However, performance improvements are limited to algorithms making heavy use of these routines.

This situation has led to the development of several contributed packages dealing with parallel computing; an overview of these efforts is provided in Schmidberger et al. (2009). **bnlearn** is designed to work with:

- The **snow** package (Tierney et al., 2008),[1] which provides support for simple parallel computing using the *master-slave* model. **snow** spawns a configurable number of R processes in background (the *slave processes*). The user can then copy data back and forth and send them commands from the R console he is working on (the *master process*). The communication between those processes is managed using either standard TCP sockets or the mechanisms provided by the **Rmpi** and **rpvm** packages. These processes are said to form a *cluster* and can run on different computers.
- The **Rmpi** package (Yu, 2010), which is an R interface to the C libraries implementing the de facto *Message-Passing Interface* (MPI) standard, a language-independent communications protocol designed to program parallel computers.
- The **rpvm** package (Li and Rossini, 2010), which is an R interface to the *Parallel Virtual Machine* (PVM) software. PVM is designed to allow a network of heterogeneous Unix and Windows machines to be used as a single distributed parallel processor.
- The **rsprng** package (Li, 2010), which provides independent random number generators to the slaves spawned by **snow**.

---

[1] Since version 2.14, the R base distribution includes a revised copy of **snow** in the **parallel** package.

We will now illustrate the use of this set of packages, which will then be used throughout this chapter to show how parallel computing applies to Bayesian networks.

The first step is to load the **snow** and the **rsprng** packages.

```
> library(snow)
> library(rsprng)
```

The **Rmpi** and **rpvm** packages are loaded by **snow** as needed. Subsequently, we need to spawn the slave processes and initialize the cluster with the makeCluster function.

```
> cl = makeCluster(2, type = "MPI")
Loading required package: Rmpi
```

The first argument of makeCluster specifies the number of slave processes which will be spawned, which is usually between 2 and the number of processes that can run concurrently without overcommitting any hardware resource. The second argument specifies the communication mechanism used between the master and the slave processes; possible values are "SOCK" to use sockets (the default), "MPI" to use **Rmpi**, and "PVM" to use **rpvm**.

Once the slave processes have been spawned, we can initialize their random number generators.

```
> clusterSetupSPRNG(cl)
```

The setup of the cluster is now completed, and we can start using it to speed up our computations. For example, we can compute simultaneously the means of all the variables of the marks data we used in Chap. 2,

```
> parApply(cl, X = marks, MARGIN = 2, FUN = mean)
    MECH     VECT      ALG      ANL     STAT
38.95455 50.59091 50.60227 46.68182 42.30682
```

getting the same result as the call to mean we would have used to compute them in a sequential way.

```
> mean(marks)
    MECH     VECT      ALG      ANL     STAT
38.95455 50.59091 50.60227 46.68182 42.30682
```

The parApply function, along with parLapply and parSapply, represents the most user-friendly way to set up embarrassingly parallel computations. These functions are the parallel versions of apply, lapply, and sapply and work in exactly the same way from the user's point of view.

Problems which are not embarrassingly parallel, or which cannot be divided in identical parts, can be tackled using a combination of clusterExport (to copy the data to the slave R processes) and clusterEvalQ (to make the slave processes execute arbitrary R commands). For instance, we may be interested in comparing Pearson's and Spearman's correlation matrices for the marks data, and we may want to estimate these matrices in parallel. To achieve that, we can first export the marks data to the slave processes,

```
> clusterExport(cl, list("marks"))
```

check that the data is now present in their global environment,

```
> unlist(clusterEvalQ(cl, ls()))
[1] "marks" "marks"
```

and then call the `cor` with the `method` argument set to `"pearson"` in the first slave process and to `"spearman"` in the second one.

```
> parLapply(cl, c("pearson", "spearman"),
+    function(m) { cor(marks, method = m) } )
[[1]]
          MECH     VECT      ALG      ANL     STAT
MECH 1.000000 0.553405 0.546751 0.409392 0.389099
VECT 0.553405 1.000000 0.609645 0.485081 0.436449
ALG  0.546751 0.609645 1.000000 0.710806 0.664736
ANL  0.409392 0.485081 0.710806 1.000000 0.607174
STAT 0.389099 0.436449 0.664736 0.607174 1.000000

[[2]]
          MECH     VECT      ALG      ANL     STAT
MECH 1.000000 0.497611 0.477201 0.415145 0.376006
VECT 0.497611 1.000000 0.609878 0.548262 0.434648
ALG  0.477201 0.609878 1.000000 0.741972 0.620780
ANL  0.415145 0.548262 0.741972 1.000000 0.628004
STAT 0.376006 0.434648 0.620780 0.628004 1.000000
```

It is also possible, even if tricky, to have each slave process execute completely different R commands. First, we need to store a unique identifier in the global environment of each slave process.

```
> slave.id = function(id) {
+    assign("id", value = id, envir = .GlobalEnv)
+ }
> parSapply(cl, 1:2, slave.id)
[1] 1 2
> clusterEvalQ(cl, id)
[[1]]
[1] 1

[[2]]
[1] 2
```

Then we can create a list, here named `calls`, containing the functions we want to call in each slave process. In this example, we consider the mean (`mean`) for the first slave and the standard deviation (`sd`) for the second one.

```
> calls = list(mean, sd)
```

It is important to note that `calls` must not have more elements than the number of slave processes referenced by the `cl` object. Otherwise, some commands will not be executed because there are more commands than slaves.

Now we can copy `calls` to the slave processes and make them execute the command stored in the element of the list that matches the identifier we stored in the `id` variable.

```
> clusterExport(cl, list("calls"))
> clusterEvalQ(cl, calls[[id]](marks))
[[1]]
    MECH      VECT      ALG      ANL      STAT
38.95455 50.59091 50.60227 46.68182 42.30682

[[2]]
    MECH      VECT      ALG      ANL      STAT
17.48622 13.14695 10.62478 14.84521 17.25559
```

The unique identifier `id` can be used in a similar way to define a list of argument sets to match the functions in `calls`, thus increasing the flexibility of the approach shown in this last example.

The synchronization of the slave processes must be managed in the master process, because slave processes cannot communicate directly with each other. Therefore, two slaves can exchange data only when the current call to `clusterEvalQ` (or `parApply`, `parSapply`, etc.) executing on the master returns, and that exchange must be taken care of by the user from the master process. This may be problematic when dealing with fine-grained parallel problems, but in most cases the performance penalty introduced by this kind of synchronization is far outweighed by the gains in performance due to the efficient use of multiple processors.

Finally, once the parallel computations are completed, we can call `stopCluster` to shut down the slave processes and update their status in the `cl` object.

```
> stopCluster(cl)
```

## 5.3 Applications to Structure and Parameter Learning

It is well known from literature that the problem of learning the structure of Bayesian networks is very hard to tackle. Its computational complexity, measured with the required number of conditional independence tests or network scores, is super-exponential in the number of nodes in the worst case and polynomial in most real-world situations. For instance, the Grow–Shrink algorithm performs a number of conditional independence test that is $O(p^2)$ for a network with $p$ nodes, under the assumption that the dimension of the Markov blanket of each node is bound by a constant. For extremely dense graphs, the number of conditional tests increases to $O(p^3 2^p)$.

Furthermore, the computational complexity of the conditional independence tests and of the network scores themselves must be taken into account; in most cases it is linear in the sample size. In R, computing conditional probabilities requires one pass over the data, while computing partial correlations requires two.

As a result, in practice most structure learning algorithms can be applied to data sets with a few hundred variables at most. Parallel implementations of learning algorithms provide a significant performance boost, thus improving our ability to handle large networks. However, it is important to note that execution time reduces at most linearly with the number of slave processes. For this reason, parallelization cannot be considered a universal solution, even though it may prove useful in many situations.

### 5.3.1 Constraint-Based Structure Learning Algorithms

Constraint-based algorithms display a coarse-grained parallelism, because they only need to synchronize their parts a couple of times. If we examine again Algorithm 2.1, we can see that:

1. The first step is embarrassingly parallel, because each d-separating set can be learned independently from the others. Another solution is to split this step in one part for each node, which will learn all the d-separating sets involving that particular node. The former approach can take advantage of a greater number of processors, while the latter has less overhead due to the smaller number of parts running in parallel.
2. The same holds for the second step. Once all the d-separating sets are known, it is embarrassingly parallel and can be split in the same way as the first step.
3. The third step is sequential, because each iteration requires the status of the previous one.

Therefore, the information available to the slave processes has to be synchronized only collected is between the first and the second step and between the second and the third step.

Most modern constraint-based algorithms, which learn the Markov blankets of the nodes as an intermediate step, require one additional synchronization. For example, if we consider the Grow–Shrink algorithm as shown in Fig. 5.1, we can see that:

1. Each Markov blanket can be computed independently from the others.
2. Each neighborhood is a subset of the corresponding Markov blanket and, therefore, can be learned independently from the others. However, the consistency of the Markov blankets must be checked before learning neighborhoods; due to errors in the conditional independence tests, they may not be symmetric (see Sect. 2.3). A solution to this problem is to examine all pairs of nodes and remove them from each other's Markov blanket if they do not appear in both of them.

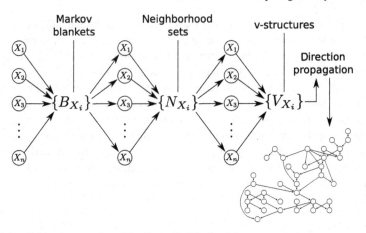

**Fig. 5.1** Parallel implementation of the Grow–Shrink algorithm present in **bnlearn**

3. Given the Markov blankets and the neighborhoods, the v-structures centered on a particular node (i.e., the one with the converging arcs) can again be identified in parallel. As in the previous step, the consistency of the neighborhoods must be checked and any departure from symmetry must be fixed beforehand.

Furthermore, the final step of the Grow–Shrink algorithm, in which the directions of compelled arcs are learned, also displays a fine-grained parallelism. The order in which arcs are considered in that step depends on the topology of the graph; undirected arcs whose orientations would result in the greatest number of cycles are considered first. That number can be computed in parallel for each arc, at the cost of introducing some overhead.

We will now examine the practical implications of parallelizing a constraint-based learning algorithm. To that end, we will use the hailfinder data set included in **bnlearn**, which is generated from the reference network of the same name. Hailfinder is a Bayesian network designed by Abramson et al. (1996) to forecast severe summer hail in northeastern Colorado. It contains 56 variables and 20,000 observations and is large enough to properly highlight the advantages and the limitations of parallel computing.

Consider a simple cluster with two slave processes.

```
> data(hailfinder)
> cl = makeCluster(2, type = "MPI")
2 slaves are spawned successfully. 0 failed.
> res = gs(hailfinder, cluster = cl)

> unlist(clusterEvalQ(cl, .test.counter))
[1]  2698 3765
> .test.counter
[1]  4
> stopCluster(cl)
```

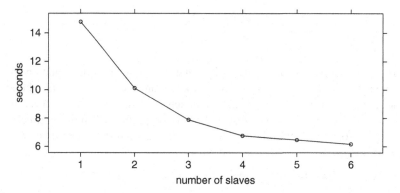

**Fig. 5.2** Performance of the Grow–Shrink algorithm for different numbers of slave processes, measured by its execution time (in seconds)

As we can see from the output of `clusterEvalQ`, the first slave process performed 2,698 (41.71 %) conditional tests, and the second one 3,765 (58.21 %). Only 4 tests were performed by the master process. The difference in the number of tests between the two slaves is due to the topology of the network; different nodes have Markov blankets and neighborhoods of different sizes, which require different numbers of tests to learn.

Increasing the number of slave processes reduces the number of tests performed by each slave, further increasing the overall performance of the algorithm.

```
> cl = makeCluster(3, type = "MPI")
3 slaves are spawned successfully. 0 failed.
> res = gs(hailfinder, cluster = cl)
> unlist(clusterEvalQ(cl, .test.counter))
[1] 1667 2198 2598
> stopCluster(cl)
> cl = makeCluster(4, type = "MPI")
4 slaves are spawned successfully. 0 failed.
> res = gs(hailfinder, cluster = cl)
> unlist(clusterEvalQ(cl, .test.counter))
[1] 1116 1582 1860 1905
> stopCluster(cl)
```

The execution times of the Grow–Shrink algorithm for clusters of 2, 3, 4, 5, and 6 slaves are reported in Fig. 5.2. It is clear from the figure that the gains in execution time follow the *law of diminishing returns*—i.e., adding more slave processes produces smaller and smaller improvements, up to the point where the increased overhead of the communications between the master and the slave processes starts actually degrading performance.

Another important consideration is whether the data set we are learning the network from actually contains enough observations and variables to make the use of the parallel implementation of a learning algorithm worthwhile. In fact, for `hailfinder` the sequential implementation of the Grow–Shrink algorithm is faster than the parallel one.

```
> system.time(gs(hailfinder))
    user   system elapsed
   4.000    0.004   4.004
```

There are three reasons for this disparity. First, the parallel implementation cannot take advantage of the symmetry of the Markov blankets and the neighborhoods to reduce the number of tests. All Markov blankets are learned simultaneously, so the same conditional independence test will likely be performed multiple times. The same holds for the neighborhoods. As a result, gs performs more than twice as many tests overall:

```
> ntests(gs(hailfinder, optimized = TRUE))
[1] 2670
> ntests(gs(hailfinder, optimized = FALSE))
[1] 6467
```

Second, tests are almost never split in an optimal way among the slave processes. This can be seen quite clearly from the examples illustrated in this section: with 4 slaves, the number of tests assigned to each of them ranges from 1,116 (17.25 % of the total) to 1,905 (29.45 % of the total). This variability introduces additional overhead in the algorithm, because slaves that have fewer tests to perform must wait for other slaves each time the status of the cluster is synchronized.

Third, passing data back and forth between the master and the slaves also takes some time. The efficiency of such an operation depends on the operating system and the hardware the cluster is running on, so it must be evaluated on a case-by-case basis.

### 5.3.2 Score-Based Structure Learning Algorithms

Score-based learning algorithms benefit from several decades of research efforts aimed at taking advantage of parallel computing in optimization heuristics.

Most score-based algorithms are inherently sequential in nature. Consider for example hill-climbing. At each iteration, the state of the previous iteration is used as the starting point for the search of a new, better network structure. This is also true for tabu search and genetic algorithms and makes the parallel implementation of these algorithms a challenging problem.

A possible solution is to provide a parallel implementation of the computations performed within a single iteration and to let the master process execute the iterations in a sequential way, synchronizing the status of the slaves each time. This would reduce a sequential problem to a fine-grained parallel one; it is known as the *move acceleration model* if each slave computes part of the score of each candidate network, or the *parallel moves model* if each slave manages some of the candidate networks. However, the resulting performance gain is likely to be outweighed by the overhead of the communications between the master and the slave processes.

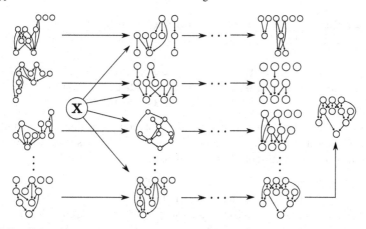

**Fig. 5.3** Parallel multistart implementation of a score-based learning algorithm

Another solution, called the *parallel multistart model*, is illustrated in Fig. 5.3 and consists in initializing several instances of a score-based algorithm with different starting networks. The use of significantly different starting points for the search improves the algorithm's ability to cover the search space and results in better and more robust solutions. For example, even if one of the instances gets stuck on a local maximum, another one may still find the global maximum. In this case, the suboptimal solution is simply discarded.

We can easily implement parallel multistart by altering the R code for model averaging used in Sect. 2.5.1. Instead of using `lapply`, we distribute the hill-climbing searches among the slave processes using `parLapply`.

```
> cl = makeCluster(4, type = "MPI")
> clusterEvalQ(cl, library(bnlearn))

> start = random.graph(names(hailfinder), num = 4, 382
+         method = "melancon")
> parallel.multistart = function(net) {
+    hc(hailfinder, start = net)
+ }
> netlist = parLapply(cl, start, parallel.multistart)
```

Once all the slave processes have completed their searches, we can evaluate the network structures they return.

```
> unlist(lapply(netlist, score, data = hailfinder))
[1] -992833.1 -993954.7 -990474.8 -1011764
```

The network with the highest score is the third one (−990474.8); the others are therefore local maxima.

Tabu search can be easily modified in the same way and with similar results.

```
> parallel.multistart = function(net) {
+    tabu(hailfinder, start = net)
+ }
> netlist = parLapply(cl, start, parallel.multistart)
> unlist(lapply(netlist, score, data = hailfinder))
[1] -990474.8 -997597.7 -991934 -993547.3
```

It is important to note that execution time is not reduced by the parallel multistart, because each of the instances executed by the slave processes takes on average as much time as the original score-based algorithm.

```
> s0 = random.graph(names(hailfinder),
+         method = "melancon")
> system.time(tabu(hailfinder, start = s0))
   user   system elapsed
414.130   0.000 414.137
> system.time(parLapply(cl, netlist,
+                            parallel.multistart))
   user   system elapsed
  0.020   0.010 432.221
```

More advanced approaches and applications are available in literature, each tailored to particular problems and with specific advantages and limitations. For an extensive coverage of such approaches, we refer the reader to Rauber and Rünger (2010).

### 5.3.3 Hybrid Structure Learning Algorithms

Applications of parallel computing to hybrid algorithms depend on the exact implementation of the *restrict* and *maximize* phases.

The restrict phase is usually implemented using the first two steps of a constraint-based algorithm or using another *local search algorithm*. Some examples of the latter are proposed in Friedman et al. (1999b) for the Sparse Candidate algorithm. More complex ones, such as ARACNE (Margolin et al., 2006), are investigated in Meloni et al. (2009). Therefore, all the considerations we made in Sect. 5.3.1 apply.

The maximize phase is usually implemented using a score-based learning algorithm. The computational cost of this phase is reduced by the constraints learned in the restrict phase, which enforce the sparseness of the network structure. This in turn guarantees a reasonable performance for most real-world data sets. All the considerations we made in Sect. 5.3.2 still apply; for example, we can still implement the multistart model if we take care to select starting networks that satisfy the constraints.

### 5.3.4 Parameter Learning

Parameter learning is another embarrassingly parallel problem. Once the structure of the network is known, the decomposition of the global distribution into the local distributions provides a natural way to split the estimation of the parameters among the slaves. The distribution of each node depends only the values of its parents and has a limited number of parameters; therefore, the amount of data copied to and from the slave processes is very small. Furthermore, assigning one variable at a time to a slave process allows an efficient use of a large number of processors.

Despite all these desirable properties, the parallel estimation of the parameters does not provide real practical advantages. First, in many "small $n$, large $p$" settings, the variables outnumber the observations. In these cases, the overhead of copying the data to the slaves is greater than the speed boost provided by parallel estimation. Second, the number of parameters is not homogeneous among the nodes. In many networks learned from biological data a small number of nodes have a large number of incoming arcs; typically they correspond to key factors in the experimental setting. Such nodes account for a large number of the parameters of the network; for example, in discrete data the number of configurations increases rapidly with the number of parents. This disparity introduces additional inefficiencies in the parallel execution, because some slaves will require much more time to complete their part of the estimation.

It is also important to note that parameter estimation is efficient in terms of computational complexity compared to most other problems concerning Bayesian networks, both in structure learning and inference. Both discrete and Gaussian Bayesian networks have closed-form estimators that can be computed in linear time (in the sample size) for the respective parameters. For this reason, the reduction in the execution time resulting from a parallel implementation is likely to be negligible over the whole analysis.

## 5.4 Applications to Inference Procedures

Inference on Bayesian networks can be performed using a variety of techniques, some specific to Bayesian networks (see Chap. 4), some defined in more general settings. Exploring applications of parallel computing to such a wide range of techniques would be impossible in the space of this chapter. For this reason, we will concentrate only on three common inference techniques: *bootstrap*, *cross-validation*, and *conditional probability queries*.

### 5.4.1 Bootstrap

Bootstrap is a very general tool for investigating probability distributions. It is also embarrassingly parallel, because bootstrap samples are mutually independent.

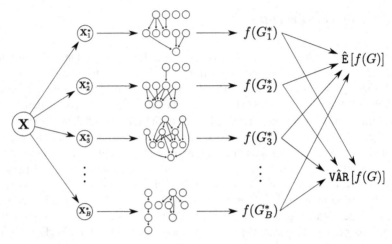

**Fig. 5.4** Nonparametric bootstrap estimate for a feature $f$ of a Bayesian network

An introduction to the relevant theory, applications, and related techniques (such as the *jackknife*) is provided in the classic monograph by Efron and Tibshirani (1993).

In Bayesian networks, bootstrap is used to investigate the properties of the parameters of the network, such as in Koller and Friedman (2009), or of its structure, such as in Friedman et al. (1999a). An illustration of the parallel implementation of such an approach is provided in Fig. 5.4. In both cases, the aspects being investigated are usually the expected value or the variance of some aspect of the Bayesian network. For example, in Friedman et al. (1999a) the statistics of interest were the probabilities associated with particular structural features of the network, such as Markov blankets or different topological orderings of the nodes. In the analysis of the Sachs et al. (2005) data covered in Sect. 2.5, the statistics of interest were the probabilities associated with each arc and its directions.

There are many other features that have a practical significance in a Bayesian network. We may be interested, for example, in the sparseness of the network we learned from the `hailfinder` data set using the hill-climbing algorithm. Sparse networks are particularly useful in analyzing real-world data: they are easier to interpret and inference is computationally tractable. We can use `bn.boot` and `narcs` to derive a point estimate and a confidence interval for the number of arcs as follows.

```
> sparse = bn.boot(hailfinder, algorithm = "hc",
+               R = 200, statistic = narcs)
> summary(unlist(sparse))
    Min. 1st Qu.  Median    Mean 3rd Qu.    Max.
   63.00   64.00   65.00   64.69   65.00   67.00
> quantile(unlist(sparse), c(0.05, 0.95))
  5% 95%
  64  66
```

`hailfinder` has 56 nodes, so with 65 arcs it can be considered sparse. Furthermore, we can see that the bootstrap estimate has a very low variance; the boundaries

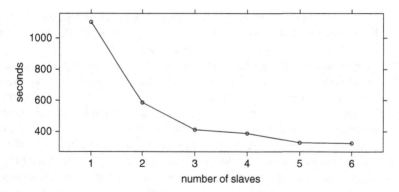

**Fig. 5.5** Performance of bootstrap resampling for different numbers of slave processes, measured by its execution time (in seconds)

of the 95 % confidence interval are very close to the mean value. This is a consequence of the large sample size of hailfinder (20,000 observations) compared to the number of parameters of the network (1,768) learned by hc.

It is easy to show that the embarrassingly parallel nature of bootstrap resampling results in substantial performance improvements:

```
> system.time(bn.boot(hailfinder, algorithm = "hc",
+     R = 200, statistic = narcs))
     user   system elapsed
1103.585    1.216 1104.848
> cl = makeCluster(2, type = "MPI")
> system.time(bn.boot(hailfinder, algorithm = "hc",
+     R = 200, statistic = narcs, cluster = cl))
   user system elapsed
  0.292   0.040 586.009
> stopCluster(cl)
```

Adding more slaves further reduces the execution time, at least up to a cluster of 6 processes (see Fig. 5.5). Using a larger number of slave processes does not result in additional speedups, at least for this number of bootstrap samples.

## 5.4.2 Cross-Validation

Cross-validation is probably the simplest and most widely used method to validate statistical models and to select suitable values for their tuning parameters. It has also been applied to many classes of models, from regression to classification, to estimate loss functions (such as *classification error* or *likelihood loss*) for model selection. Several examples of such applications are covered in Hastie et al. (2009).

Similarly to the bootstrap, cross-validation is embarrassingly parallel. Once the data have been partitioned in $k$ parts and the $k$ cross-validation samples $\mathbf{X}^*_{-1}, \ldots, \mathbf{X}^*_{-k}$

have been created, a Bayesian network is learned in parallel, independently, from each split. The corresponding losses can also be computed in parallel and then averaged to produce a cross-validated loss estimate (see Fig. 5.6).

Most Bayesian network structure learning algorithms are not explicitly targeted at classification problems; they seek to minimize the discrepancy between the estimated and the true dependence structure rather than classification error. Furthermore, the very concept of a target variable is central in classification but alien to Bayesian networks, which treat all the variables in the same way. However, there are some situations in which the classification error, estimated with the *prediction error*, may be of interest. For example, the Hailfinder network was designed to forecast severe summer hail in northeastern Colorado. In fact, the nodes whose names end in Fcst represent the weather conditions in different parts of the region, and the prediction of their values was the main goal of the original work by Abramson et al. (1996).

If we focus on CompPlFcst (*Complete Plains Forecast*), we can see that Max-Min Hill-Climbing is not able to learn a good classifier.

```
> bn.cv(hailfinder, 'mmhc', loss = "pred",
+    loss.args = list(target = "CompPlFcst"))

  k-fold cross-validation for Bayesian networks

  target learning algorithm:
                                  Max-Min Hill-Climbing
  number of subsets:                      10
  loss function:
                                  Classification Error
  expected loss:                        0.5433
```

**Fig. 5.6** K-fold cross-validation estimation of a loss function for a Bayesian network learning algorithm

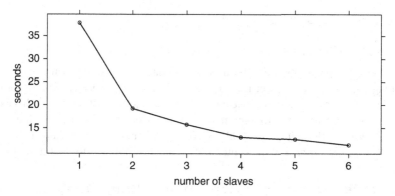

**Fig. 5.7** Performance of cross-validation for different numbers of slave processes, measured by its execution time (in seconds)

Hill-climbing and tabu search have comparable error rates (50.67 % and 50.4 %, respectively). As was the case for the bootstrap, using two slave processes halves the execution time.

```
> system.time(bn.cv(hailfinder, 'mmhc', loss = "pred",
+    loss.args = list(target = "CompPlFcst")))
   user   system elapsed
 37.782    0.056  37.836
> cl = makeCluster(2, type = "MPI")

> system.time(bn.cv(hailfinder, 'mmhc', loss = "pred",
+    loss.args = list(target = "CompPlFcst"),
+    cluster = cl))
   user   system elapsed
  0.196    0.032  19.202
> stopCluster(cl)
```

Adding more slaves to the cluster improves the performance of the cross-validation further, at least up to cluster of size 6 (see Fig. 5.7).

```
> cl = makeCluster(3, type = "MPI")
> system.time(bn.cv(hailfinder, 'mmhc', loss = "pred",
+    loss.args = list(target = "CompPlFcst"),
+    cluster = cl))
   user   system elapsed
  0.320    0.016  15.726
> stopCluster(cl)
> cl = makeCluster(4, type = "MPI")
> system.time(bn.cv(hailfinder, 'mmhc', loss = "pred",
+    loss.args = list(target = "CompPlFcst"),
+    cluster = cl))
```

```
      user   system elapsed
     0.440    0.036  13.007
> stopCluster(cl)
```

In fact, cross-validation can be used to evaluate any combination of structure learning algorithms, parameter learning methods, and the respective tuning parameters. It can also be used to evaluate a predetermined network structure; in this case, $\mathbf{X}^*_{-1}, \ldots, \mathbf{X}^*_{-k}$ are used only for parameter learning. Consider, for instance, the *naive Bayes classifier* (Borgelt et al., 2009), which is equivalent to a star-shaped network with the training variable at the center and all the arcs pointing to the training variable.

```
> naive = naive.bayes(training = "CompPlFcst",
+    data = hailfinder)
> bn.cv(hailfinder, naive, loss = "pred")


  k-fold cross-validation for Bayesian networks


  target network structure:
   [Naive Bayes Classifier]
  number of subsets:                        10
  loss function:
                            Classification Error
  training node:                      CompPlFcst
  expected loss:                      0
```

As expected, the classification error is considerably lower than with hc or tabu. Naive Bayes is, despite its simple structure and strong assumptions, one of the most efficient and effective algorithms in data mining and classification (Zhang, 2004).

The performance gain from the use of a **snow** cluster is not as marked as in the previous example (the execution time halves with two slaves, but does not improve beyond that). This difference in behavior suggests that most of the execution time in the previous example was spent learning the structure of the network and that, as anticipated in Sect. 5.3.4, parameter learning is relatively fast in comparison.


## 5.4.3 Conditional Probability Queries

Conditional probability queries are the most common form of Bayesian network inference; as a result, parallel implementations of the exact and approximate algorithms covered in Sect. 4.1.2 have been investigated in literature. Particle filters algorithms, in particular, exhibit coarse-grained parallelism if particles are generated using Markov chain Monte Carlo approaches or are embarrassingly parallel if particles are independent. Logic sampling, illustrated in Algorithm 4.2, falls in the second category.

Consider, for example, how the knowledge that there is a weather instability in the mountains (i.e., InsInMt == "Strong") and that there is a marked cloud

shading (i.e., `CldShadeConv == "Marked"`) influences Hailfinder's forecasts
for the plains. To investigate this influence, we use logic sampling to generate $10^7$
observations from the Bayesian network learned by mmhc (`hailfinder`).

```
> fitted = bn.fit(mmhc(hailfinder), hailfinder)
> n = nrow(hailfinder)

> summary(hailfinder[, "CompPlFcst"]) / n
DecCapIncIns IncCapDecIns LittleChange
     0.22810      0.41205      0.35985
> cp = cpdist(fitted, nodes = "CompPlFcst",
+ (InsInMt == "Strong") & (CldShadeConv == "Marked"),
+    n = 10^7)
> n = nrow(cp)
> summary(cp[, CompPlFcst]) / n
DecCapIncIns IncCapDecIns LittleChange
   0.1888219    0.4812025    0.3299755
```

The three levels of `CompPlFcst` stand for *decreased instability* (`DecCap
IncIns`), *increased instability* (`IncCapDecIns`), and *little change* (`Little
Change`). The conditional distribution shows an increased probability of the
weather worsening ($+6.9\%$) compared to the marginal one, which suggests that bad
weather tends to spill from the mountains into the plains. This trend is confirmed
by the decreased probability of `DecCapIncIns` ($-3.9\%$) and `LittleChange`
($-2.9\%$).

The embarrassingly parallel nature of logic sampling results in very noticeable
performance gains even with only 2 slave processes; execution times for up to 6
slaves are shown in Fig. 5.8.

```
> system.time(cpdist(fitted, nodes = "CompPlFcst",
+ (InsInMt == "Strong") & (CldShadeConv == "Marked"),
+    n = 10^7, batch = 10^6))
```

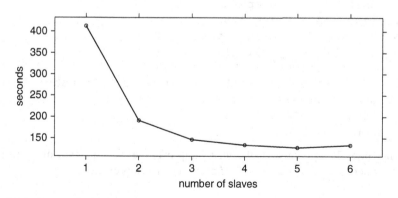

**Fig. 5.8** Performance of conditional probability queries for different numbers of slave processes,
measured by its execution time (in seconds)

```
    user   system elapsed
385.632   26.798 412.549
> cl = makeCluster(2, type = "MPI")
> system.time(cpdist(fitted, nodes = "CompPlFcst",
+ (InsInMt == "Strong") & (CldShadeConv == "Marked"),
+   n = 10^7, cluster = cl))
    user   system elapsed
   8.713   0.244 191.079
> stopCluster(cl)
```

The application of logic sampling to the estimation of conditional probabilities, instead of whole distributions, shows similar performance improvements. Consider, for example, the probability that the wind is blowing toward the west in the mountains (i.e., WindFieldMt == "Westerly") conditional to the fact that it is blowing from east/northeast in the plains (i.e., WindFieldPln == "E_NE").

```
> cpquery(fitted, (WindFieldMt == "Westerly"),
+   (WindFieldPln == "E_NE"), n = 10^7)
[1] 0.4136172
> n = nrow(hailfinder)

> summary(hailfinder[, "WindFieldMt"]) / n
LVorOther   Westerly
   0.47615    0.52385
```

The conditional probability is lower than the marginal one because the plains and the mountains are adjacent and winds cannot completely change in direction so suddenly.

Using two slave processes again halves the execution time.

```
> system.time(cpquery(fitted,
+    (WindFieldMt == "Westerly"),
+    (WindFieldPln == "E_NE"), n = 10^7, batch = 10^6))
    user   system elapsed
291.439   14.889 306.328
> cl = makeCluster(2, type = "MPI")
> system.time(cpquery(fitted,
+    (WindFieldMt == "Westerly"),
+    (WindFieldPln == "E_NE"), n = 10^7, cluster = cl))
    user   system elapsed
   0.004    0.004 178.921
> stopCluster(cl)
```

Adding more slave processes to the cluster improves the execution time of cpquery even further, as was the case for cpdist in the previous example.

## Exercises

**5.1.** Using the `hailfinder` data set included in **bnlearn** and a **snow** cluster with at least 2 slave processes:

(a) Compute the number of levels and the most common level for each node.
(b) Split the samples among the slaves and identify which nodes have at least one level with less than 5 observations in that particular subsample.
(c) Compute the entropy of each variable in `hailfinder`, defined as

$$H(\mathbf{p}) = \sum -p \log p,$$

where $p$ is the relative frequency of each level of the variable.

**5.2.** Consider the `alarm` data set included in **bnlearn**.

(a) Learn the structure of the network using Inter-IAMB and a shrinkage test with `alpha = 0.01` and measure the execution time of the algorithm.
(b) Does a 2-node cluster provide a greater performance improvement than just switching from `optimized = FALSE` to `optimized = TRUE`?
(c) Is that still true when a Monte Carlo permutation test is used?

**5.3.** Consider again the `alarm` data set from Exercise 5.2, and a **snow** cluster with at least 2 nodes.

(a) Use nonparametric bootstrap to determine the distribution of the number of arcs present in a network structure learned with hc.
(b) How does that distribution change when bootstrap samples have size m = 100?
(c) Compare the distribution of the number of score comparisons for m = 100 and m = 5000.

**5.4.** Implement a parallel version of the model averaging performed using hc with random starting networks in Sect. 2.5.1.

# Solutions

## Exercises of Chap. 1

**1.1 Consider a directed acyclic graph with $n$ nodes.**

**(a) Show that at least one node must not have any incoming arc, i.e., the graph must contain at least one root node.**

**(b) Show that such a graph can have at most $\frac{1}{2}n(n-1)$ arcs.**

**(c) Show that a path can span at most $n-1$ arcs.**

**(d) Describe an algorithm to determine the topological ordering of the graph.**

(a) Let $G = (\mathbf{V}, A)$ be a directed graph in which each node has at least one outgoing arc. This is equivalent to saying that each node has at least one incoming arc; therefore, $G$ does not have any root node. Choose a vertex $v_1 \in \mathbf{V}$. Since $v_1$ has an outgoing arc, there is a vertex $v_2$ such that $v_1 \to v_2$. Proceeding in this manner, we obtain a path $v_1 \to \ldots \to v_n$ spanning all the nodes in $G$. But $v_n$ also has an outgoing arc; therefore, the graph is not acyclic.

(b) Choose a vertex $v_1 \in \mathbf{V}$. $v_1$ can have at most $n-1$ outgoing arcs. Consider now a second node $v_2 \neq v_1$. $v_2$ can have at most $n-2$ outgoing arcs; the arc $v_2 \to v_1$ would create a cycle and is therefore disregarded. By induction, we have that

$$|A| \leqslant (n-1) + (n-2) + \ldots + (1) = \binom{n}{2} = \frac{n(n-1)}{2}.$$

(c) Suppose, by contradiction, that there exists a path spanning $n$ arcs. A path can (by definition) pass at most once through each node, and the path is spanning $n+1$ nodes. Therefore, the path is passing twice through at least one node, implying that the path is in fact a cycle. This contradicts the assumption that the graph is acyclic.

(d) Two simple ways of determining the topological ordering of a directed acyclic graph are the *breadth-first search* and the *depth-first search* algorithms, described in Bang-Jensen and Gutin (2009) and Russell and Norvig (2009).

R. Nagarajan et al., *Bayesian Networks in R: with Applications in Systems Biology*, Use R! 48, DOI 10.1007/978-1-4614-6446-4, © Springer Science+Business Media New York 2013

**1.2  Consider the graphs shown in Fig. 1.1.**

**(a) Obtain the skeleton of the partially directed and directed graphs.**

**(b) Enumerate the acyclic graphs that can be obtained by orienting the undirected arcs of the partially directed graph.**

**(c) List the arcs that can be reversed (i.e., turned in the opposite direction), one at a time, without introducing cycles in the directed graph.**

(a) The skeletons of the partially directed and directed graphs are, respectively:

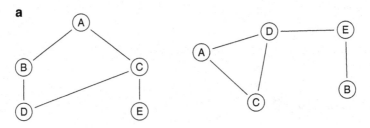

(b) Only six acyclic orientations of the partially directed graph are possible:

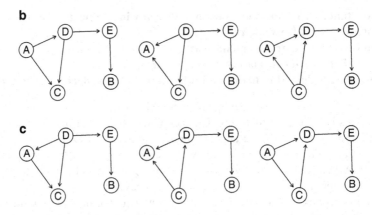

(c) All arcs of the directed graph can be reversed without introducing cycles.

**1.3  The (famous) iris data set reports the measurements in centimeters of the sepal length and width and the petal length and width for 50 flowers from each of 3 species of iris ("setosa," "versicolor," and "virginica").**

**(a) Load the iris data set (it is included in the data set package, which is part of the base R distribution and does not need to be loaded explicitly) and read its manual page.**

**(b) Investigate the structure of the data set.**

**(c) Compare the sepal length among the three species by plotting histograms side by side.**

**(d) Repeat the previous point using boxplots.**

```
(a) > data(iris)
    > ?iris
```

```
(b) > summary(iris)
    > dim(iris)
(c) > par(mfrow = c(1, 3))
    > hist(iris[iris[, "Species"] == "setosa",
    +                          "Sepal.Length"],
    +    xlab = "length", main = "Setosa Sepal Length")
    > hist(iris[iris[, "Species"] == "versicolor",
    +                          "Sepal.Length"],
    +    xlab = "length",
    +    main = "Versicolor Sepal Length")
    > hist(iris[iris[, "Species"] == "virginica",
    +                          "Sepal.Length"],
    +    xlab = "length",
    +    main = "Virginica Sepal Length")
(d) > boxplot(Sepal.Length ~ Species, data = iris)
```

**1.4 Consider again the iris data set from Exercise 1.3.**

(a) Write the data frame holding iris data frame into a space-separated text file named "iris.txt", and read it back into a second data frame called iris2.

(b) Check that iris and iris2 are identical.

(c) Repeat the previous two steps with a file compressed with bzip2 named "iris.txt.bz2".

(d) Save iris directly (e.g., without converting it to a text table) into a file called "iris.rda", and read it back.

(e) List all R objects in the global environment, and remove all of them apart from iris.

(f) Exit the R saving the contents of the current session.

```
(a) > write.table(iris, file = "iris.txt",
    +    row.names = FALSE)
    > iris2 = read.table("iris.txt", header = TRUE)
(b) > identical(iris, iris2)
(c) > bzfd = bzfile("iris.txt.bz2", open = "w")
    > write.table(iris, file = bzfd, row.names = FALSE)
    > close(bzfd)
    > bzfd = bzfile("iris.txt.bz2", open = "r")
    > iris2 = read.table(bzfd, header = TRUE)
    > close(bzfd)
    > identical(iris, iris2)
(d) > save(iris, file = "iris.rda")
    > load("iris.rda")
(e) > ls()
    > l = ls()
    > rm(list = l[l != "iris"])
```

(f) > quit (save = "yes")

**1.5  Consider the `gaussian.test` data set included in bnlearn.**

(a) **Print the column names.**
(b) **Print the range and the quartiles of each variable.**
(c) **Print all the observations for which A falls in the interval $[3,4]$ and B in**
    $(-\infty, -5] \cup [10, \infty)$.
(d) **Sample 50 rows without replacement.**
(e) **Draw a bootstrap sample (e.g., sample 5000 observations with replacement)**
    **and compute the mean of each variable.**
(f) **Standardize each variable.**

```
(a) > colnames (gaussian.test)
    > names (gaussian.test)
(b) > for (var in names (gaussian.test))
    +    print (range (gaussian.test [, var]))
    > for (var in names (gaussian.test))
    +    print (quantile (gaussian.test [, var],
    +       probs = (1:3)/4))
(c) > condA = (gaussian.test [, "A"] >= 3) &
    +                        (gaussian.test [, "A"] <= 4)
    > condB = (gaussian.test [, "B"] <= -4) |
    +                        (gaussian.test [, "B"] >= 4)
    > gaussian.test [condA & condB, ]
(d) > gaussian.test [sample (50, replace = FALSE), ]
(e) > colMeans (gaussian.test [
    +    sample (5000, replace = TRUE), ])
(f) > scale (gaussian.test)
```

**1.6  Generate a data frame with $100$ observations for the following variables:**

(a) **A categorical variable with two levels, `low` and `high`. The first $50$ obser-**
    **vations should be set to `low`, the others to `high`.**
(b) **A categorical variable with two levels, `good` and `bad`, nested within the**
    **first variable, i.e., the first $25$ observations should be set to `good`, the second**
    **$25$ to `bad`, and so on.**
(c) **A continuous, numerical variable following a Gaussian distribution with**
    **mean $2$ and variance $4$ when the first variable is equal to `low` and with**
    **mean $4$ and variance $1$ if the first variable is equal to `high`.**

**In addition, compute the standard deviation of the last variable for each con-**
**figuration of the first two variables.** The variables can be generated as follows:

```
(a) > A = factor (c (rep ("low", 50), rep ("high", 50)),
    +          levels = c ("low", "high"))
(b) > nesting = c (rep ("good", 25), rep ("bad", 25))
    > B = factor (rep (nesting, 2),
    +          levels = c ("good", "bad"))
```

(c) > C = c(rnorm(50, mean = 2, sd = 2),
    +          rnorm(50, mean = 4, sd = 1))

and the data frame can then be created with

    > data = data.frame(A = A, B = B, C = C)

Then, the standard deviations can be computes as

    > by(data[, "C"], INDICES = data[, c("A", "B")],
    +    FUN = sd)


## Exercises of Chap. 2

**2.1** Consider the `asia` synthetic data set from Lauritzen and Spiegelhalter (1988), which describes the diagnosis of a patient at a chest clinic who has just come back from a trip to Asia and is showing dyspnea.

(a) Load the data set from the bnlearn package and investigate its characteristics using the exploratory analysis techniques covered in Chap. 1.
(b) Create a `bn` object with the network structure described in the manual page of `asia`.
(c) Derive the skeleton, the moral graph, and the CPDAG representing the equivalence class of the network. Plot them using `graphviz.plot`.
(d) Identify the parents, the children, the neighbors, and the Markov blanket of each node.

(a) > summary(asia)
    > dim(asia)
(b) > spec = "[A] [S] [T|A] [L|S] [B|S] [D|B:E] [E|T:L] [X|E]"
    > bn = model2network(spec)
(c) > bn.skel = skeleton(bn)
    > graphviz.plot(bn.skel)
    > bn.moral = moral(bn)
    > graphviz.plot(bn.moral)
    > bn.eq = cpdag(bn)
    > graphviz.plot(bn.eq)
(d) > sapply(nodes(bn), parents, x = bn)
    > sapply(nodes(bn), children, x = bn)
    > sapply(nodes(bn), nbr, x = bn)
    > sapply(nodes(bn), mb, x = bn)

**2.2** Using the network structures created in Exercise 2.1 for the `asia` data set, produce the following plots with `graphviz.plot`:

(a) A plot of the CPDAG of the equivalence class in which the arcs belonging to a v-structure are highlighted (either with a different color or using a thicker line width).

**(b)** Fill the nodes with different colors according to their role in the diagnostic process: causes ("visit to Asia" and "smoking"), effects ("tuberculosis," "lung cancer," and "bronchitis"), and the diagnosis proper ("chest X-ray," "dyspnea," and "either tuberculosis or lung cancer/bronchitis").

**(c)** Explore different layouts by changing the `layout` and `shape` arguments.

```
(a) > vs = vstructs(bn.eq, arcs = TRUE)
    > graphviz.plot(bn.eq, highlight =
    +    list(arcs = vs, lwd = 2, col = "grey"))
(b) > graphviz.plot(bn.eq,
    +    highlight = list(nodes = nodes(bn),
    +    fill = c("blue", "red", "green", "green", "red",
    +       "blue", "red", "green"), col = "black"))
(c) > par(mfrow = c(2, 5))
    > layout = c("dot", "neato", "twopi", "circo",
      "fdp")
    > shape = c("ellipse", "circle")
    > for (l in layout) {
    +    for (s in shape) {
    +       main = paste(l, s)
    +       graphviz.plot(bn.eq, shape = s, layout = l,
    +          main = main)
    +    }
    + }
```

**2.3** Consider the `marks` data set analyzed in Sect. 2.3.

**(a)** Discretize the data using a quantile transform and different numbers of intervals (say, from 2 to 5). How does the network structure learned from the resulting data sets change as the number of intervals increases?

**(b)** Repeat the discretization using interval discretization using up to 5 intervals, and compare the resulting networks with the ones obtained previously with quantile discretization.

**(c)** Does Hartemink's discretization algorithm perform better than either quantile or interval discretization? How does its behavior depend on the number of initial breaks?

**(a)** As the number of intervals increases, fewer and fewer arcs are included in the network. This is a consequence of the loss of information resulting from discretizing variables one at a time, without considering their joint distribution.

```
> intervals = 2:5
> par(mfrow = c(1, length(intervals)))
> for (int in intervals) {
+    dmarks = discretize(marks, breaks = int,
+                method = "quantile")
+    main = paste("dmarks,", int, "intervals")
```

```
+    graphviz.plot(hc(dmarks), main = main)
+ }
```

(b) Interval discretization's performance is comparable with the one of quantile discretization. Again, as the number of intervals increases, the dependence relationships linking the variables are lost, and fewer and fewer arcs are picked up by the structure learning algorithm.

```
> intervals = 2:5
> par(mfrow = c(1, length(intervals)))
> for (int in intervals) {
+    dmarks = discretize(marks, breaks = int,
+                method = "interval")
+    main = paste("dmarks,", int, "intervals")
+    graphviz.plot(hc(dmarks), main = main)
+ }
```

(c) Hartemink's discretization performs better than both interval and quantile discretization if we start with a suitably high number of breaks (ibreaks = 50) and perform the initial discretization with a quantile transform (idisch = "interval"). Even when each variable is discretized into 4 intervals, a substantial part of the network structure is still learned correctly.

```
> intervals = 2:5
> par(mfrow = c(1, length(intervals)))
> for (int in intervals) {
+    dmarks = discretize(marks, breaks = int,
+                method = "hartemink", ibreaks = 50,
+                idisc = "interval")
+    main = paste("dmarks,", int, "intervals")
+    graphviz.plot(hc(dmarks), main = main)
+ }
```

**2.4 The ALARM network (Beinlich et al. 1989) is a Bayesian network designed to provide an alarm message system for patients hospitalized in intensive care units (ICU). Since ALARM is commonly used as a benchmark in literature, a synthetic data set of 5000 observations generated from this network is available from bnlearn as alarm.**

(a) **Create a bn object for the "true" structure of the network using the model string provided in its manual page.**
(b) **Compare the networks learned with different constraint-based algorithms with the true one, both in terms of structural differences and using either BIC or BDe.**
(c) **The overall performance of constraint-based algorithms suggests that the asymptotic $\chi^2$ conditional independence tests may not be appropriate for analyzing alarm. Are permutation or shrinkage tests better choices?**
(d) **How are the above learning strategies affected by changes to alpha?**

(a) The model string reported in the manual page is the following:

```
> true = empty.graph(names(alarm))
> modelstring(true) = paste("[HIST|LVF] [CVP|LVV]",
+    "[PCWP|LVV] [HYP] [LVV|HYP:LVF] [LVF]",
+    "[STKV|HYP:LVF] [ERLO] [HRBP|ERLO:HR]",
+    "[HREK|ERCA:HR] [ERCA] [HRSA|ERCA:HR] [ANES]",
+    "[APL] [TPR|APL] [ECO2|ACO2:VLNG] [KINK]",
+    "[MINV|INT:VLNG] [FIO2] [PVS|FIO2:VALV]",
+    "[SAO2|PVS:SHNT] [PAP|PMB] [PMB] [SHNT|INT:PMB]",
+    "[INT] [PRSS|INT:KINK:VTUB] [DISC] [MVS]",
+    "[VMCH|MVS] [VTUB|DISC:VMCH]",
+    "[VLNG|INT:KINK:VTUB] [VALV|INT:VLNG]",
+    "[ACO2|VALV] [CCHL|ACO2:ANES:SAO2:TPR]",
+    "[HR|CCHL] [CO|HR:STKV] [BP|CO:TPR]", sep = "")
```

(b) The performance of constraint learning algorithms for the alarm data set improves in newer algorithms, which is not unexpected because this network is used so often as a benchmark in publications. However, note how with the default parameters the network structures that are learned from the data differ from true. The overall number of arcs in each network is similar to the number of arcs in true, but many false positives and false negatives are present. As a result, network scores for the learned networks are much lower than the one for the true network.

```
> bn.gs = gs(alarm)
> bn.iamb = iamb(alarm)
> bn.inter = inter.iamb(alarm)
> par(mfrow = c(2, 2))
> graphviz.plot(true, main = "True Structure")
> graphviz.plot(bn.gs, main = "Grow-Shrink")
> graphviz.plot(bn.iamb, main = "IAMB")
> graphviz.plot(bn.inter, main = "Inter-IAMB")
> unlist(compare(true, bn.gs))
> unlist(compare(true, bn.iamb))
> unlist(compare(true, bn.inter))
> score(cextend(bn.gs), alarm, type = "bde")
> score(cextend(bn.iamb), alarm, type = "bde")
> score(cextend(bn.inter), alarm, type = "bde")
```

(c) Permutation tests significantly improve the performance of both IAMB and Inter-IAMB: true positives (i.e., arcs correctly included in the network) increase for all learning algorithms. False positives (i.e., arcs incorrectly included in the network) and false negatives (i.e., arcs incorrectly absent from the network) are also fewer, though by a smaller amount.

```
> bn.gs2 = gs(alarm, test = "smc-x2")
> bn.iamb2 = iamb(alarm, test = "smc-x2")
> bn.inter2 = inter.iamb(alarm, test = "smc-x2")
> unlist(compare(true, bn.gs2))
> unlist(compare(true, bn.iamb2))
> unlist(compare(true, bn.inter2))
```

(d) Shrinkage tests improve the results of structure learning much like permutation tests, but with a much smaller execution time.

```
> bn.gs3 = gs(alarm, test = "smc-x2", B = 10000,
>               alpha = 0.01)
> bn.iamb3 = iamb(alarm, test = "smc-x2",
>               B = 10000, alpha = 0.01)
> bn.inter3 = inter.iamb(alarm, test = "smc-x2",
>               B = 10000, alpha = 0.01)
> unlist(compare(true, bn.gs3))
> unlist(compare(true, bn.iamb3))
> unlist(compare(true, bn.inter3))
```

**2.5  Consider again the alarm network used in Exercise 2.4.**

**(a) Learn its structure with hill-climbing and tabu search, using the posterior density BDe as a score function. How does the network structure change with the imaginary sample size iss?**

**(b) Does the length of the tabu list have a significant impact on the network structures learned with tabu?**

**(c) How does the BIC score compare with BDe at different sample sizes in terms of structure and score of the learned network?**

(a) In both hill-climbing and tabu search the number of arcs increases with the value of iss. Since the imaginary sample size determines how much weight is assigned to the prior distribution compared to the sample, it also controls the amount of smoothing applied to the posterior density. For this reason, (comparatively) large values of iss oversmooth the data and result in widely different network having similar scores and, in turn, allow too many arcs to be included in the network.

```
> par(mfrow = c(2, 5))
> for (iss in c(1, 5, 10, 20, 50)) {
+     bn = hc(alarm, score = "bde", iss = iss)
+     main = paste("hc(..., iss = ", iss, ")",
+               sep = "")
+     sub = paste(narcs(bn), "arcs")
+     graphviz.plot(bn, main = main, sub = sub)
+ }
> for (iss in c(1, 5, 10, 20, 50)) {
+     bn = tabu(alarm, score = "bde", iss = iss)
```

```
+    main = paste("tabu(..., iss = ", iss, ")",
+              sep = "")
+    sub = paste(narcs(bn), "arcs")
+    graphviz.plot(bn, main = main, sub = sub)
+  }
```

(b) The length of the tabu list does have a significant impact on structure learning, for two reasons. First of all, it does increase the number of network structures that are by `tabu`, and therefore structure learning requires more time. This is especially relevant for score functions that are expensive to compute, such as BGe. Furthermore, the score of network structure consistently increases with the length of the tabu list; getting stuck into a local maximum becomes more and more unlikely as the tabu list grows.

```
> par(mfrow = c(1, 5))
> for (n in c(10, 15, 20, 50, 100)) {
+    bn = tabu(alarm, score = "bde", tabu = n)
+    bde = score(bn, alarm, type = "bde")
+    main = paste("tabu(..., tabu = ", n, ")",
+              sep = "")
+    sub = paste(ntests(bn), "steps, score", bde)
+    graphviz.plot(bn, main = main, sub = sub)
+  }
```

(c) The BIC score is asymptotically equivalent to BDe, so the networks learned using these two scores become more similar as sample size increases. At small sample sizes, BIC penalizes dense networks more heavily than BDe and therefore results in much fewer arcs being included and in much lower execution time.

```
> par(mfrow = c(2, 6))
> for (n in c(100, 200, 500, 1000, 2000, 5000)) {
+    bn.bde = hc(alarm[1:n, ], score = "bde")
+    bn.bic = hc(alarm[1:n, ], score = "bic")
+    bde = score(bn.bde, alarm, type = "bde")
+    bic = score(bn.bic, alarm, type = "bic")
+    main = paste("BDe, sample size", n)
+    sub = paste(ntests(bn.bde), "steps, score", bde)
+    graphviz.plot(bn.bde, main = main, sub = sub)
+    main = paste("BIC, sample size", n)
+    sub = paste(ntests(bn.bic), "steps, score", bic)
+    graphviz.plot(bn.bic, main = main, sub = sub)
+  }
```

**2.6   Consider the observational data set from Sachs et al. (2005) used in Sect. 2.5.1 (the original data set, not the discretized one).**

(a) **Evaluate the networks learned by hill-climbing with BIC and BGe using cross-validation and the log-likelihood loss function.**

**(b)** Use bootstrap resampling to evaluate the distribution of the number of arcs present in each of the networks learned in the previous point. Do they differ significantly?

**(c)** Compute the averaged network structure for `sachs` using hill-climbing with BGe and different imaginary sample sizes. How does the value of the significance threshold change as `iss` increases?

(a) The network learned with BGe appears to fit the data better than the one fitted with BIC, but not by a wide margin. Therefore, we need to repeat cross-validation for a suitable number of times to conclude the difference is significant.

```
> sachs = read.table("sachs.data.txt",
+          header = TRUE)
> bn.bic = hc(sachs, score = "bic-g")
> bn.cv(bn.bic, data = sachs)
> bn.bge = hc(sachs, score = "bge")
> bn.cv(bn.bge, data = sachs)
```

(b) The distributions of the number of arcs for BIC and BGe present important differences. First, the latter is bell-shaped, while the former is markedly asymmetric. Second, the mean and the standard deviations of the two distributions are different (the exact values depend on the bootstrap samples, so they change at each new simulation).

```
> narcs.bic =
+    bn.boot(sachs, algorithm = "hc",
+            algorithm.args = list(score = "bic-g"),
+            statistic = narcs)
> narcs.bge =
+    bn.boot(sachs, algorithm = "hc",
+            algorithm.args = list(score = "bge"),
+            statistic = narcs)
> narcs.bic = unlist(narcs.bic)
> narcs.bge = unlist(narcs.bge)
> par(mfrow = c(1, 2))
> hist(narcs.bic, main = "BIC", freq = FALSE)
> curve(dnorm(x, mean = mean(narcs.bic),
+    sd = sd(narcs.bic)), add = TRUE, col = 2)
> hist(narcs.bge, main = "BGe", freq = FALSE)
> curve(dnorm(x, mean = mean(narcs.bge),
+    sd = sd(narcs.bge)), add = TRUE, col = 2)
```

(c) 
```
> t = numeric(5)
> iss = c(5, 10, 20, 50, 100)
> for (i in seq_along(iss)) {
+    s = boot.strength(sachs, algorithm = "hc",
+            algorithm.args = list(score = "bge",
```

```
   +                                                         iss = iss[i]]))
   +    t[i] = attr(s, "threshold")
   +  }
```

## Exercises of Chap. 3

**3.1** Consider the `Canada` data set from the vars package, which we analyzed in Sect. 3.5.1.

(a) Load the data set from the vars package and investigate its properties using the exploratory analysis techniques covered in Chap. 1.
(b) Estimate a VAR(1) process for this data set.
(c) Build the auto-regressive matrix $A$ and the constant matrix $B$ defining the VAR(1) model.
(d) Compare the results with the LASSO matrix when estimating the $L_1$-penalty with cross-validation.
(e) What can you conclude ?

```
(a) > data(Canada)
    > summary(Canada)
(b) > var.1c = VAR(Canada, p = 1, type = "const")
(c) > coefficients = coef(var.1c)
    > mat = matrix(0, 4, 5)
    > pvalue = 0.05
    > pos = which(coefficients$e[, "Pr(>|t|)"] < pvalue)
    > mat[1, pos] = coefficients$e[pos, "Estimate"]
    > pos =
    +  which(coefficients$prod[, "Pr(>|t|)"] < pvalue)
    > mat[2, pos] = coefficients$prod[pos, "Estimate"]
    > pos =
    +  which(coefficients$rw[, "Pr(>|t|)"] < pvalue)
    > mat[3, pos] = coefficients$rw[pos, "Estimate"]
    > pos = which(coefficients$U[, "Pr(>|t|)"] < pvalue)
    > mat[4, pos] = coefficients$U[pos, "Estimate"]
    > A = mat[, 1:4]
    > B = matrix(mat[, 5], 4, 1)
(d) > library(lars)
    > data = Canada
    > x = data[-c(dim(data)[1]), ]
    > fit.all = lapply(colnames(data),
    +    function(gene) {
    +      y = data[-c(1), gene]
    +      lars(y = y, x = x, type = "lasso")
    +  })
```

```
> cv.pred.all = lapply(1:dim(data)[2],
+   function(gene) {
+       y = data[-c(1), gene]
+       lasso.cv = cv.lars(y = y, x = x,
+                   mode = "fraction")
+       frac = lasso.cv$index[which.min(lasso.cv$cv)]
+       predict(fit.all[[gene]], s = frac,
+           type = "coef", mode = "fraction")
+   })
> cv.pred.all[[1]]$coefficients
> cv.pred.all[[2]]$coefficients
> cv.pred.all[[3]]$coefficients
> cv.pred.all[[4]]$coefficients
```

(e) We can conclude that the LASSO is not selective enough when there are too few variables. In this case (4 variables and 18 time points), the classic VAR process inference procedure provided by the **vars** package is more appropriate.

**3.2   Consider the `arth800` data set from the GeneNet package, which we analyzed in Sects. 3.5.2 and 3.5.3.**

(a) **Load the data set from the GeneNet package. The time series expression of the 800 genes is included in a data set called `arth800.expr`. Investigate its properties using the exploratory analysis techniques covered in Chap. 1.**
(b) **For this practical exercise, we will work on a subset of variables (one for each gene) having a large variance. Compute the variance of each of the 800 variables, plot the various variance values in decreasing order, and create a data set with the variables greater than 2.**
(c) **Can you fit a VAR process with a usual approach from this data set?**
(d) **Which alternative approaches can be used to fit a VAR process from this data set?**
(e) **Estimate a dynamic Bayesian network with each of the alternative approaches presented in this chapter.**

(a)
```
> library(GeneNet)
> data(arth800)
> summary(arth800.expr)
> dim(arth800.expr)
```
The data contains 2 sets of 11 time points.

(b)
```
> variance = diag(var(arth800.expr))
> plot(sort(variance, decreasing = TRUE),
+   type = "l", ylab = "Variance")
> abline(h = 2, lty = 2)
> posVar2 = which(variance > 2)
> dataVar2 = arth800.expr[, posVar2]
> dim(dataVar2)
```

(c) It is not possible to fit a VAR process with the default approach proposed in the
package **vars** as the number of variable is greater (49 genes) than the number
of measurements (22 time points).
If we try to do that,

```
> dataVar2inline = dataVar2[c(seq(1, 22, by = 2),
+   seq(2, 22, by = 2)), ]
> library(vars)
> var.1c = VAR(data, p = 1, type = "const")
```

The estimated coefficient contains many missing values (NA); therefore, approaches allowing for dimension reduction are required to analyze these data.

(d) We consider the following dimension reduction approaches:

- $L_1$ norm penalty (LASSO)
- James-Stein shrinkage
- Low-order conditional dependencies approximation

(e) Various approaches for reduction dimension

- LASSO with **lars**:

```
> library(lars)
> data = dataVar2inline
> x = data[-c(21:22), ]
> fit.all = lapply(colnames(data),
+   function(gene) {
+     y = data[-(1:2), gene]
+     lars(y = y, x = x, type = "lasso")
+   })
> cv.pred.all = lapply(1:dim(data)[2],
+   function(gene) {
+     y = data[-(1:2), gene]
+     lasso.cv = cv.lars(y = y, x = x,
+                   mode = "fraction")
+     frac = lasso.cv$index[
+               which.min(lasso.cv$cv)]
+     predict(fit.all[[gene]], s = frac,
+        type = "coef", mode = "fraction")
+   })
> DBNlasso = matrix(0, dim(data)[2], dim(data)[2])
> for (i in 1:dim(DBNlasso)[1]) {
+   DBNlasso[i, ] =
+     cv.pred.all[i][[1]]$coefficients
+ }
> # percentage of arcs
> sum(DBNlasso != 0)
[1] 421
```

```
> sum(DBNlasso != 0)/prod(dim(DBNlasso))
[1] 0.1753436
> plot(sort(abs(DBNlasso), decr = TRUE)[1:500],
+   type = "l",
+   ylab = "Absolute coefficients")
```

- James-Stein shrinkage with **GeneNet**:

```
> DBNGeneNet = ggm.estimate.pcor(dataVar2,
+                   method = "dynamic")
> # p-values, q-values and posterior probabilities
> # for each potential arc
> DBNGeneNet.edges =
+  network.test.edges(DBNGeneNet)
> # plot the arcs probability by decreasing order
> plot(DBNGeneNet.edges[, "prob"], type = "l")
> # number of arcs with prob > 0.9
> # (i.e. local fdr < 0.1)
> sum(DBNGeneNet.edges$prob > 0.9)
```

- First-order conditional dependencies approximation with **G1DBN**:

```
> library(G1DBN)
> data = dataVar2inline
> # next step is a bit long but is less than
> # 3 minutes with a regular PC
> DBNG1DBNstep1 =
+  DBNScoreStep1(data, method = "ls")
> DBNG1DBN = DBNScoreStep2(DBNG1DBNstep1$S1ls,
+               data, method = "ls", alpha1 = 0.5)
> plot(sort(DBNG1DBN), type = "l",
+   ylab = "Arcs' p-values")
```

- LASSO and network modularity with **SIMoNe**:

```
> library(simone)
> data = dataVar2inline
> ctrl = setOptions(clusters.crit = "BIC")
> DBNsimone.BIC =
+  simone(data, type = "time-course",
+    clustering = TRUE, control = ctrl)
> DBNsimone.BIC.net = getNetwork(DBNsimone.BIC)
> # number of arcs:
> sum(DBNsimone.BIC.net$A == 1)
```

**3.3  Consider the dimension reduction approaches used in the previous exercise and the arth800 data set from the GeneNet package.**

**(a)** **For a comparative analysis of the different approaches, select the top** 50 **arcs for each approach (function** `BuildEdges` **from the G1DBN package can be used to that end).**

**(b)** **Plot the four inferred networks with the function** `plot` **from package G1DBN.**

**(c)** **How many arcs are common to the four inferred networks?**

**(d)** **Are the top** 50 **arcs of each inferred network similar? What can you conclude?**

**Using the results of the previous exercise.**

(a) Top 50 arcs computation.

- With the LASSO:

```
> lasso.thres.top50 = mean(sort(abs(DBNlasso),
+                          decreasing = TRUE)[50:51])
> DBNlasso.50edges =
+    BuildEdges(score = -abs(DBNlasso),
+        threshold = -lasso.thres.top50)
```

- With **GeneNet**:

```
> DBNGeneNet.50edges =
+    cbind(DBNGeneNet.edges[1:50, "node1"],
+          DBNGeneNet.edges[1:50, "node2"])
```

- With **SIMoNe**:

```
> nbArcs = 50
> ctrl = setOptions(clusters.crit = nbArcs)
> DBNsimone.50 = simone(data,type = "time-course",
+                clustering = TRUE,control = ctrl)
> DBNsimone.50.net = getNetwork(DBNsimone.50,
+                          selection = nbArcs)
> DBNsimone.edges =
+   BuildEdges(score = -DBNsimone.50.net$A,
+        threshold = 0)
```

- With **G1DBN**:

```
> G1DBN.thres.top50 = mean(sort(DBNG1DBN)[50:51])
> DBNG1DBN.edges =
+    BuildEdges(score = DBNG1DBN,
+        threshold = G1DBN.thres.top50, prec = 3)
```

(b) Plots of the top 50 arcs identified by each approach.

- With the LASSO:

```
> par(mfrow = c(2, 2))
> DBNlasso.top50 =
```

```
+     graph.edgelist(cbind(DBNlasso.50edges[, 1],
+       DBNlasso.50edges[, 2]))
> DBNlasso.nodeCoord =
+     layout.fruchterman.reingold(DBNlasso.top50)
> plot(DBNlasso.top50,
+     layout = DBNlasso.nodeCoord,
+     edge.arrow.size = 0.5, vertex.size = 10,
+     main = "Network inferred with LASSO"")
```

- With **GeneNet**:

```
> DBNGeneNet.top50 =
+   graph.edgelist(DBNGeneNet.50edges)
> DBNGeneNet.nodeCoord =
+ layout.fruchterman.reingold(DBNGeneNet.50edges)
> plot(DBNGeneNet.top50,
+   layout = DBNGeneNet.nodeCoord,
+   edge.arrow.size = 0.5, vertex.size = 10,
+   main = "Network inferred with GeneNet")
```

- With **SIMoNe**:

```
> DBNsimone.top50 =
+     graph.edgelist(cbind(DBNsimone.50edges[, 1],
+                         DBNsimone.50edges[, 2]))
> DBNsimone.nodeCoord =
+     layout.fruchterman.reingold(DBNsimone.top50)
> plot(DBNsimone.top50,
+     layout = DBNsimone.nodeCoord,
+     edge.arrow.size = 0.5, vertex.size = 10,
+     main = "Network inferred with SIMoNe")
```

- With **G1DBN**:

```
> DBNG1DBN.top50 =
+     graph.edgelist(cbind(DBNG1DBN.50edges[, 1],
+                         DBNG1DBN.50edges[, 2]))
> DBNG1DBN.nodeCoord =
+     layout.fruchterman.reingold(DBNG1DBN.top50)
> plot(DBNG1DBN.top50, layout=DBNG1DBN.nodeCoord,
+     edge.arrow.size = 0.5, vertex.size = 10,
+     main = "Network inferred with G1DBN")
```

(c) Arcs common to all 4 inferred networks:

```
> DBNlasso.50edges.mat =
+   as.numeric(abs(DBNlasso) > lasso.thres.top50)
> DBNGeneNet.50edges.mat = matrix(0, 49, 49)
```

```
> for (i in 1:50){
>   DBNGeneNet.50edges.mat[DBNGeneNet.edges$
    node2[i],
+     DBNGeneNet.edges$node1[i]] = 1
> }#FOR
> DBNsimone.50edges.mat = DBNsimone.50edges.net$A
> DBNG1DBN.50edges.mat =
+   as.numeric(DBNG1DBN < G1DBN.thres.top50)
> sum(DBNG1DBN.50edges.mat, na.rm = TRUE)
> sum(which(DBNlasso.50edges.mat == 1) %in%
+   which(DBNGeneNet.50edges.mat == 1))
> sum(which(DBNlasso.50edges.mat == 1) %in%
+   which(DBNG1DBN.50edges.mat == 1))
> sum(which(DBNlasso.50edges.mat == 1) %in%
+   which(DBNsimone.50edges.mat == 1))
```

(d) Different dimension reduction procedures select significantly different sets of
arcs. It is likely that the various approaches have different power in identifying
the dependencies present in the data and therefore complement each other.

## Exercises of Chap. 4

**4.1 Apply the junction tree algorithm to the validated network structure from
Sachs et al. (2005), and draw the resulting undirected triangulated graph.**

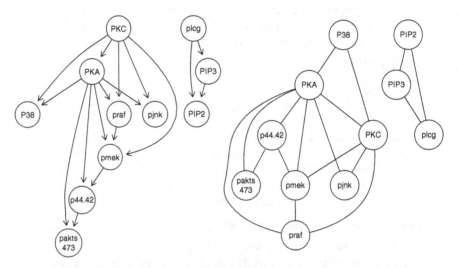

**4.2 Consider the Sachs et al. (2005) data used in Sect. 4.2.**

(a) **Perform parameter learning with the bn.fit function from bnlearn and
the validated network structure. How do the maximum likelihood estimates**

differ from the Bayesian ones, and how do the latter vary as the imaginary sample size increases?

(b) Node **PKA** is parent of all the nodes in the **praf** → **pmek** → **p44.42** → **pakts473** chain. Use the junction tree algorithm to explore how our beliefs on those nodes change when we have evidence that PKA is "LOW," and when PKA is "HIGH."

(c) Similarly, explore the effects on **pjnk** of evidence on **PIP2**, **PIP3**, and **plcg**.

(a) Parameter fitting can be carried out as follows:

```
> library(bnlearn)
> val.spec = paste("[PKC][PKA|PKC][praf|PKC:PKA]",
+       "[pmek|PKC:PKA:praf][p44.42|pmek:PKA]",
+       "[pakts473|p44.42:PKA][P38|PKC:PKA]",
+       "[pjnk|PKC:PKA][plcg][PIP3|plcg]",
+       "[PIP2|plcg:PIP3]")
> val = model2network(val.spec)
> isachs = isachs[, 1:11]
> for (i in names(isachs))
+    levels(isachs[, i]) = c("LOW", "AVG", "HIGH")
> fitted = bn.fit(val, isachs, method = "mle")
> fitted2 = bn.fit(val, isachs, method = "bayes",
+                 iss = 5)
> fitted3 = bn.fit(val, isachs, method = "bayes",
+                 iss = 10)
> fitted4 = bn.fit(val, isachs, method = "bayes",
+                 iss = 20)
```

The main difference between maximum likelihood and Bayesian estimates is that the former can contain NaNs, while the latter are always completely specified. This happens when, for a particular node, some of the parents' configurations are not observed; therefore, the distribution of the node conditional on that configuration cannot be estimated from the data. In Bayesian estimates, such unknown conditional distributions result in the posterior being equal to the prior, that is, uniform.

As for the effects of the imaginary sample size, larger values result in smoother estimates because of the increasing weight given to the uniform prior compared to the sample. Posterior parameter estimates move away from 0 and 1, and therefore inference is affected by fewer numerical problems.

(b) Continuing from the previous point,

```
> library(gRain)
> jtree = compile(as.grain(fitted))
> jprop = setFinding(jtree, nodes = "PKA",
+              states = "LOW")
> query = c("praf", "pmek", "p44.42", "pakts473")
```

```
> querygrain(jtree, nodes = query)
> querygrain(jprop, nodes = query)
> jprop = setFinding(jtree, nodes = "PKA",
+               states = "HIGH")
> querygrain(jtree, nodes = query)
> querygrain(jprop, nodes = query)
```

When PKA is HIGH, the activity of all the proteins corresponding to the query nodes is inhibited (the LOW probability increases and the HIGH decreases). When PKA is LOW, the opposite is true (the LOW probability decreases and the HIGH increases).

(c) Continuing from the previous two points,

```
> jprop = setFinding(jtree,
+               nodes = c("PIP2", "PIP3", "plcg"),
+               states = rep("LOW", 3))
> a = querygrain(jtree, nodes = "pjnk")
> b = querygrain(jprop, nodes = "pjnk")
> identical(a, b)
```

Our belief on the pjnk node is completely unaffected by any evidence on either PIP2, PIP3, or plcg, because there is no path from the former to the any of the latter. Therefore, changes in belief due to new evidence cannot propagate from PIP2, PIP3, and plcg to pjnk.

**4.3  Consider the marks data set analyzed in Sect. 2.3.**

**(a) Learn both the network structure and the parameters with likelihood-based approaches, i.e., BIC or AIC, for structure learning and maximum likelihood estimates for the parameters.**

**(b) Query the network learned in the previous point for the probability to have the marks for both STAT and MECH above 60, given evidence that the mark for ALG is at most 60. Are the two variables independent given the evidence on ALG?**

**(c) What is the (conditional) probability of having an average vote (in the [60,70] range) in both VECT and MECH while having an outstanding vote in ALG (at least 90)?**

(a) 
```
> bn = hc(marks, score = "bic-g")
> fitted = bn.fit(bn, marks)
```
(b) 
```
> cpquery(fitted,
+    event = (STAT > 60) & (MECH > 60),
+    evidence = (ALG <= 60), n = 5 * 10^6)
> cpquery(fitted, event = (STAT > 60),
+    evidence = (ALG <= 60), n = 5 * 10^6)
> cpquery(fitted, event = (MECH > 60),
+    evidence = (ALG <= 60), n = 5 * 10^6)
```

The conditional probability of the two events is not equal to the product of the corresponding marginal probabilities; therefore, STAT and MECH are not independent given the evidence on ALG. Note that the fact that ALG d-separates STAT and MECH is not relevant in this case, because the evidence on ALG is soft evidence (i.e., ALG is still a random variable, just with a different distribution).

(c) 
```
> cpquery(fitted,
+    event = ((MECH >= 60) & (MECH <= 70)) |
+            ((VECT >= 60) & (VECT <= 70)),
+    evidence = (ALG >= 90), n = 5 * 10^6)
```

**4.4  Using the dynamic Bayesian network dbn2 from Sect. 4.3, investigate the effects of genes 257710_at and 255070_at observed at time $t-2$ on gene 265768_at at time $t$.**

```
> cpquery(dbn2.fit, event = ('265768_at' > 8),
+                   evidence = ('257710_at' > 8))
[1] 0.3571429
> cpquery(dbn2.fit, event = ('265768_at' > 8),
+                   evidence = ('255070_at' > 8))
[1] 0.5903756
> cpquery(dbn2.fit, event = ('265768_at' > 8),
+                   evidence = TRUE)
[1] 0.4427231
```

High expression levels of gene 257710_at at time $t-2$ reduce the probability of high expression levels of gene 265768_at at time $t$; the opposite is true for gene 255070_at.

## Exercises of Chap. 5

**5.1  Using the hailfinder data set included in bnlearn and a snow cluster with at least 2 slave processes:**

(a) **Compute the number of levels and the most common level for each node.**
(b) **Split the samples among the slaves, and identify which nodes have at least one level with less than 5 observations in that particular subsample.**
(c) **Compute the entropy of each variable in hailfinder, defined as**

$$H(\mathbf{p}) = \sum -p \log p$$

**where $p$ is the relative frequency of each level of the variable.**

(a) 
```
> library(bnlearn)
> library(snow)
> clusterSetupSPRNG(cl)
> cl = makeCluster(2, type = "SOCK")
```

```
> parSapply(hailfinder, cl = cl, FUN = nlevels)
> most.common = function(x) {
+     names(which.max(table(x)))
+ }#MOST.COMMON
> parSapply(cl, hailfinder, most.common)
```
(b)
```
> folds = split(sample(nrow(hailfinder)),
+               seq_len(length(cl)))
> small.counts = function(x, data) {
+     sapply(data[x, ], function(y) any(table(y) <= 5))
+ }
> parSapply(cl, folds, small.counts,
+     data = hailfinder)
```
(c)
```
> h = function(x) {
+     p = prop.table(table(x))
+     return(sum(-p*log(p)))
> }#H
> parSapply(cl, hailfinder, h)
> stopCluster(cl)
```

**5.2  Consider the `alarm` data set included in bnlearn.**

(a) **Learn the structure of the network using Inter-IAMB and a shrinkage test with `alpha` = 0.01, and measure the execution time of the algorithm.**

(b) **Does a 2-node cluster provide a greater performance improvement than just switching from `optimized` = `FALSE` to `optimized` = `TRUE`?**

(c) **Is that still true when a Monte Carlo permutation test is used?**

(a)
```
> library(snow)
> cl = makeCluster(2, type = "SOCK")
> clusterSetupSPRNG(cl)
> system.time(inter.iamb(alarm, test = "mi-sh",
+     alpha = 0.01))
```
(b)
```
> system.time(inter.iamb(alarm, test = "mi-sh",
+     alpha = 0.01, optimized = FALSE))
> system.time(inter.iamb(alarm, test = "mi-sh",
+     alpha = 0.01, cluster = cl))
```
The nonparallelized, optimized algorithm is about twice as fast as the non-optimized one. It is also faster than using a 2-node cluster by about 30%.

(c)
```
> system.time(inter.iamb(alarm, test = "mc-mi",
+     alpha = 0.01, B = 1000))
> system.time(inter.iamb(alarm, test = "mc-mi",
+     alpha = 0.01, B = 1000, optimized = FALSE))
> system.time(inter.iamb(alarm, test = "mc-mi",
+     alpha = 0.01, B = 1000, cluster = cl))
> stopCluster(cl)
```

In this case, the difference in execution time between the parallel and the optimized versions of Inter-IAMB is much smaller, because the extra time required by permutation tests (compared to shrinkage ones) makes the overhead of the **snow** cluster much less noticeable.

**5.3  Consider again the `alarm` data set from Exercise 5.2, and a snow cluster with at least 2 nodes.**

**(a) Use nonparametric bootstrap to determine the distribution of the number of arcs present in a network structure learned with hc.**
**(b) How does that distribution change when bootstrap samples have size m = 100?**
**(c) Compare the distribution of the number of score comparisons for m = 100 and m = 5000.**

(a) ```
> library(snow)
> cl = makeCluster(2, type = "SOCK")
> clusterSetupSPRNG(cl)
> n = bn.boot(alarm, narcs, R = 50,
+         algorithm = "hc", cluster = cl)
> hist(unlist(n))
```
The distribution of the number of arcs is very tight around 55. It is skewed to the left and varies in the range $[53, 58]$. This suggests that the sample size of `alarm` is large enough to reliably learn the structure of the network.

(b) ```
> n = bn.boot(alarm, narcs, R = 50, m = 1000
+         algorithm = "hc", cluster = cl)
> hist(unlist(n))
```
The distribution of the number of arcs has a much greater spread the one studied in the previous point; it varies in the range $[24, 36]$. Its expectation is also much smaller than before. Furthermore, the distribution is not as skewed as before.

(c) ```
> n1 = bn.boot(alarm, ntests, R = 50, m = 100,
>         algorithm = "hc", cluster = cl)
> n2 = bn.boot(alarm, ntests, R = 50, m = 5000,
>         algorithm = "hc", cluster = cl)
> par(mfrow = c(1, 2))
> hist(unlist(n1), main = "m = 100")
> hist(unlist(n2), main = "m = 5000")
> stopCluster(cl)
```
The number of network scores computed for m = 5000 is greater than the corresponding number for m = 100; the former varies in the range $[2466, 2610]$, while the latter in $[1602, 1854]$. Both distributions are skewed on the left and have similar spreads.

**5.4  Implement a parallel version of the model averaging performed using hc with random starting networks in Sect. 2.5.1.**
```

```
> library(snow)
> cl = makeCluster(2, type = "SOCK")
> clusterSetupSPRNG(cl)
> clusterEvalQ(cl, library(bnlearn))
> sachs = read.table("code/sachs.data.txt",
+           header = TRUE)
> dsachs = discretize(sachs, method = "hartemink",
+           breaks = 3, ibreaks = 60)
> clusterExport(cl, list("dsachs"))
>  nodes = names(dsachs)
>  start = random.graph(nodes = nodes,
+           method = "melancon", num = 50)
>  netlist = parLapply(cl, start, function(net) {
+    hc(dsachs, score = "bde", iss = 10, start = net)})
> rnd = custom.strength(netlist, nodes = nodes)
> rnd[(rnd$strength > 0.85) & (rnd$direction >= 0.5), ]
> avg.start = averaged.network(rnd, threshold = 0.85)
> stopCluster(cl)
```

# References

Abramson B, Brown J, Edwards W, Murphy A, Winkler RL (1996) Hailfinder: a Bayesian system for forecasting severe weather. Int J Forecast 12(1):57–71

Aliferis CF, Statnikov A, Tsamardinos I, Mani S, Xenofon XD (2010a) Local causal and Markov blanket induction for causal discovery and feature selection for classification part I: algorithms and empirical evaluation. J Mach Learn Res 11:171–234

Aliferis CF, Statnikov A, Tsamardinos I, Mani S, Xenofon XD (2010b) Local causal and Markov blanket induction for causal discovery and feature selection for classification part II: analysis and extensions. J Mach Learn Res 11:235–284

Balov N (2011) mugnet: mixture of Gaussian Bayesian network model. R package version 0.13.5

Balov N, Salzman P (2012) catnet: categorical Bayesian Network inference. R package version 1.13.4

Bang-Jensen J, Gutin G (2009) Digraphs: theory, algorithms and applications, 2nd edn. Springer, Heidelberg

Beal M, Falciani F, Ghahramani Z, Rangel C, Wild D (2005) A Bayesian approach to reconstructing genetic regulatory networks with hidden factors. Bioinformatics 21:349–356

Beinlich IA, Suermondt HJ, Chavez RM, Cooper GF (1989) The ALARM monitoring system: a case study with two probabilistic inference techniques for belief networks. In: Proceedings of the 2nd European conference on artificial intelligence in medicine, Springer, pp 247–256

Bera AK, Jarque CM (1981) Efficient tests for normality, homoscedasticity and serial independence of regression residuals: Monte Carlo evidence. Econ Lett 7(4):313–318

Borgelt C, Steinbrecher M, Krus R (2009) Graphical models: representations for learning, reasoning and data mining, 2nd edn. Wiley, New York

Bøttcher SG, Dethlefsen C (2003) deal: a package for learning Bayesian networks. J Stat Softw 8(20):1–40

Bouckaert RR (1995) Bayesian belief networks: from construction to inference. PhD thesis, Utrecht University, The Netherlands

Castelo R, Roverato A (2006) A robust procedure for Gaussian graphical model search from microarray data with p larger than n. J Mach Learn Res 7:2621–2650

Castillo E, Gutiérrez JM, Hadi AS (1997) Expert systems and probabilistic network models. Springer, New York

Cheng J, Druzdel MJ (2000) AIS-BN: an adaptive importance sampling algorithm for evidential reasoning in large Bayesian networks. J Artif Intell Res 13:155–188

Chickering DM (1995) A transformational characterization of equivalent Bayesian network structures. In: Proceedings of the 11th conference on uncertainty in artificial intelligence (UAI95), pp 87–98

Chickering DM (1996) Learning Bayesian networks is NP-complete. In: Fisher D, Lenz H (eds) Learning from data: artificial intelligence and statistics V. Springer, New York, pp 121–130

R. Nagarajan et al., *Bayesian Networks in R: with Applications in Systems Biology*,
Use R! 48, DOI 10.1007/978-1-4614-6446-4,
© Springer Science+Business Media New York 2013

Chiquet J, Smith A, Grasseau G, Matias C, Ambroise C (2009) SIMoNe: statistical inference for modular networks. Bioinformatics 25(3):417–418

Claeskens G, Hjort NL (2008) Model selection and model averaging. Cambridge University Press, Cambridge

Cooper GF (1990) The computational complexity of probabilistic inference using Bayesian belief networks. Artif Intell 42(2–3):393–405

Cooper GF, Yoo C (1995) Causal discovery from a mixture of experimental and observational data. In: UAI '99: Proceedings of the 15th annual conference on uncertainty in artificial intelligence, Morgan Kaufmann, pp 116–125

Cover TM, Thomas JA (2006) Elements of information theory, 2nd edn. Wiley, Hoboken

Csardi G, Nepusz T (2006) The igraph software package for complex network research. Int J Comp Syst:1695, pp 1–38

Diestel R (2005) Graph theory, 3rd edn. Springer, Heidelberg

Dor D, Tarsi M (1992) A simple algorithm to construct a consistent extension of a partially oriented graph. Technical Report, UCLA, Cognitive Systems Laboratory, available as Technical Report R-185

Dondelinger F, Lèbre S, Husmeier D (2012) EDISON: Software for network reconstruction and changepoint detection. R package version 1.0

Dondelinger F, Lèbre S, Husmeier D (2013) Non-homogeneous dynamic Bayesian networks with Bayesian regularization for inferring gene regulatory networks with gradually time-varying structure. Machine Learning 90(2):191–230

Edwards DI (2000) Introduction to graphical modelling, 2nd edn. Springer, New York

Efron B, Tibshirani R (1993) An introduction to the bootstrap. Chapman & Hall, New York

Efron B, Hastie T, Johnstone I, Tibshirani R (2004) Least angle regression. Ann Stat 32(2):407–499

Engle RF (1982) Autoregressive conditional heteroscedasticity with estimates of the variance of United Kingdom inflation. Econometrica 50(4):987–1007

Fienberg SE (1980) The analysis of cross-classified categorical data, 2nd edn. Springer, New York

Flynn MJ (1972) Some computer organizations and their effectiveness. IEEE Trans Comput 21(9):948–960

Friedman N, Goldszmidt M (1996) Discretizing continuous attributes while learning Bayesian networks. In: Proceedings of the 13th international conference on machine learning (ICML96), Morgan Kaufmann

Friedman N, Goldszmidt M, Wyner A (1999a) Data analysis with Bayesian networks: a bootstrap approach. In: Proceedings of the 15th conference on uncertainty in artificial intelligence, pp 196–205

Friedman J, Hastie T, Tibshirani R (2010) Regularization paths for generalized linear models via coordinate descent. J Stat Softw 33(1):1–22

Friedman N, Linial M, Nachman I, Pe'er D (2000) Using Bayesian network to analyze expression data. J Comput Biol 7:601–620

Friedman N, Pe'er D, Nachman I (1999b) Learning Bayesian network structure from massive datasets: the "Sparse Candidate" algorithm. In: Proceedings of 15th conference on uncertainty in artificial intelligence (UAI), Morgan Kaufmann, pp 206–215

Geiger D, Heckerman D (1994) Learning Gaussian networks. Technical Report, Microsoft Research, Redmond, Washington, available as Technical Report MSR-TR-94-10

Gentleman R, Whalen E, Huber W, Falcon S (2012) graph: a package to handle graph data structures. R package version 1.32.0

Gentry J, Long L, Gentleman R, Seth, Hahne F, Sarkar D, Hansen K (2012) Rgraphviz: provides plotting capabilities for R graph objects. R package version 1.32.0

Goeman JJ (2012) penalized R package. R package version 0.9-41

Green PJ (1995) Reversible jump Markov chain Monte Carlo computation and Bayesian model determination. Biometrika 82(4):711–732

Hamilton JD (1994) Time-series analysis. Princeton University Press, Princeton

Hartemink AJ (2001) Principled computational methods for the validation and discovery of genetic regulatory networks. PhD thesis, School of Electrical Engineering and Computer Science, Massachusetts Institute of Technology

Hastie T, Efron B (2012) lars: least angle regression, lasso and forward stagewise. R package version 1.1

Hastie T, Tibshirani R, Friedman J (2009) The elements of statistical learning: data mining, inference, and prediction, 2nd edn. Springer, New York

Hausser J, Strimmer K (2009) Entropy inference and the James-Stein estimator, with application to nonlinear gene association networks. J Mach Learn Res 10:1469–1484

Heckerman D, Geiger D, Chickering DM (1995) Learning Bayesian networks: the combination of knowledge and statistical data. Mach Learn, 20(3):197–243. Available as Technical Report MSR-TR-94-09.

Højsgaard S (2010) gRain: graphical independence networks. R package version 0.8.5

Højsgaard S, Dethlefsen C, Bowsher C (2010) gRbase: a package for graphical modelling in R. R package version 1.3.4

Ide JS, Cozman FG (2002) Random generation of Bayesian networks. In: SBIA '02: Proceedings of the 16th Brazilian symposium on artificial intelligence, Springer, pp 366–375

Imoto S, Goto T, Miyano S (2002) Estimation of genetic networks and functional structures between genes by using Bayesian networks and nonparametric regression. In: Proceedings of the 7th Pacific symposium on biocomputing, pp 175–186

Imoto S, Kim S, Goto T, Aburatani S, Tashiro K, Kuhara S, Miyano S (2003) Bayesian network and nonparametric heteroscedastic regression for nonlinear modeling of genetic network. J Bioinforma Comput Biol 2:231–252

James W, Stein C (1961) Estimation with quadratic loss. In: Neyman J (ed) Proceedings of the 4th Berkeley symposium on mathematical statistics and probability, pp 361–379

Jarque CM, Bera AK (1980) Efficient tests for normality, homoscedasticity and serial independence of regression residuals. Econ Lett 6(3):255–259

Jarque CM, Bera AK (1987) A test for normality of observations and regression residuals. Int Stat Rev 55(2):163–172

Jensen FV (2001) Bayesian networks and decision graphs. Springer, New York

Kalisch M, Mächler M, Colombo D, Maathuis MH, Bühlmann P (2012) Causal inference using graphical models with the R package pcalg. J Stat Softw 47(11):1–26

Kim S, Imoto S, Miyano S (2003) Inferring gene networks from time series microarray data using dynamic Bayesian networks. Brief Bioinform 4(3):228

Kim S, Imoto S, Miyano S (2004) Dynamic Bayesian network and nonparametric regression for nonlinear modeling of gene networks from time series gene expression data. Biosystems 75(1–3):57–65

Kohavi R, Sahami M (1996) Error-based and entropy-based discretization of continuous features. In: Proceedings of the 2nd international conference on knowledge discovery and data mining (KDD '96), AAAI Press, pp 114–119

Koller D, Friedman N (2009) Probabilistic graphical models: principles and techniques. MIT Press, Cambridge

Korb KB, Hope LR, Nicholson AE, Axnick K (2004) Varieties of causal intervention. In: Zhang C, Guesgen HW, Yeap W (eds) Proceedings of 8th Pacific rim international conference on artificial intelligence (PRICAI 2004), Springer, pp 322–331

Korb KB, Nicholson AE (2010) Bayesian artificial intelligence 2nd edn. Chapman and Hall, Boca Raton

Kullback S (1968) Information theory and statistic. Dover Publications, New York

Kulinskaya E, Morgenthaler S, Staudte RG (2008) Meta analysis: a guide to calibrating and combining statistical evidence. Wiley, Hoboken

Larrañaga P, Sierra B, Gallego MJ, Michelena MJ, Picaza JM (1997) Learning Bayesian networks by genetic algorithms: a case study in the prediction of survival in malignant skin melanoma. In: Proceedings of the 6th conference on artificial intelligence in medicine in Europe (AIME '97), Springer, pp 261–272

Lauritzen SL (1996) Graphical models. Oxford University Press, Oxford

Lauritzen SL, Spiegelhalter DJ (1988) Local computation with probabilities on graphical structures and their application to expert systems (with discussion). J R Stat Soc Series B Stat Methodol 50(2):157–224

Lèbre S (2008) G1DBN: a package performing dynamic Bayesian network inference. R package version 3.1

Lèbre S (2009) Inferring dynamic genetic networks with low order independencies. Stat Appl Genet Mol Biol 8(1):9

Lèbre S, Becq J, Devaux F, Lelandais G, Stumpf M (2010) Statistical inference of the time-varying structure of gene-regulation networks. BMC Syst Biol 4(130):1–16

Ledoit O, Wolf M (2003) Improved estimation of the covariance matrix of stock returns with an application to portfolio selection. J Empir Financ 10:603–621

Legendre P (2000) Comparison of permutation methods for the partial correlation and partial mantel tests. J Stat Comput Simul 67:37–73

Li NM (2010) rsprng: R interface to SPRNG (Scalable Parallel Random Number Generators). R package version 1.0

Li NM, Rossini AJ (2010) rpvm: R interface to PVM (Parallel Virtual Machine). R package version 1.0-4

Lütkepohl H (2005) New introduction to multiple time series analysis. Springer, New York

MacLachlan GJ, Krishnan T (2008) The EM algorithm and extensions, 2nd edn. Wiley, Hoboken

Mardia KV, Kent JT, Bibby JM (1979) Multivariate analysis. Academic, London

Margaritis D (2003) Learning Bayesian network model structure from data. PhD thesis, School of Computer Science, Carnegie-Mellon University, Pittsburgh, PA, available as Technical Report CMU-CS-03-153

Margolin A, Nemenman I, Basso K, Wiggins C, Stolovitzky G, Favera R, Califano A (2006) ARACNE: an algorithm for the reconstruction of gene regulatory networks in a mammalian cellular context. BMC Bioinformatics 7(Suppl 1):S7

Meinshausen N, Bühlman P (2006) High dimensional graphs and variable selection with the LASSO. Ann Stat 34(3):1436–1462

Melançon G, Dutour I, Bousquet-Mélou M (2001) Random generation of directed acyclic graphs. Electronic Notes Discrete Math 10:202–207

Meloni A, Ripoli A, Positano V, Landini L (2009) Improved learning of Bayesian networks in biomedicine. In: Proceedings of the 9th international conference on intelligent systems design and applications, IEEE Computer Society, pp 624–628

Murphy KP (2002) Dynamic Bayesian networks: representation, inference and learning. PhD thesis, Computer Science Division, UC Berkeley

Neapolitan RE (2003) Learning Bayesian networks. Prentice Hall, Englewood Cliffs

Ong IM, Glasner JD, Page D (2002) Modelling regulatory pathways in E. Coli from time series expression profiles. Bioinformatics 18(Suppl 1):S241–S248

Opgen-Rhein R, Strimmer K (2007) Learning causal networks from systems biology time course data: an effective model selection procedure for the vector autoregressive process. BMC Bioinformatics 8(Suppl. 2):S3

Pearl J (1988) Probabilistic reasoning in intelligent systems: networks of plausible inference. Morgan Kaufmann, Los Altos

Pearl J (2009) Causality: models, reasoning and inference, 2nd edn. Cambridge University Press, Cambridge

Perrin BE, Ralaivola L, Mazurie A, Bottani S, Mallet J, d'Alché Buc F (2003) Gene networks inference using dynamic Bayesian networks. Bioinformatics 19(Suppl 2):S138–S148

Pfaff B (2008a) Analysis of integrated and cointegrated time series with R, 2nd edn. Springer, New York

Pfaff B (2008b) VAR, SVAR and SVEC models: implementation within R package vars. J Stat Softw 27(4):1–32

R Development Core Team (2012) R: a language and environment for statistical computing. R Foundation for Statistical Computing, Vienna, Austria, URL http://www.R-project.org, ISBN 3-900051-07-0

Rangel C, Angus J, Ghahramani Z, Lioumi M, Sotheran E, Gaiba A, Wild DL, Falciani F (2004) Modeling T-cell activation using gene expression profiling and state-space models. Bioinformatics 20(9):1361–1372

Rauber T, Rünger G (2010) Parallel programming for multicore and cluster systems. Springer, Berlin

Rissanen J (2007) Information and complexity in statistical models. Springer, New York

Robert CP, Casella G (2009) Introducing Monte Carlo methods with R. Springer, New York

Russell SJ, Norvig P (2009) Artificial intelligence: a modern approach, 3rd edn. Prentice Hall, Englewood Cliffs

Sachs K, Perez O, Pe'er D, Lauffenburger DA, Nolan GP (2005) Causal protein-signaling networks derived from multiparameter single-cell data. Science 308(5721):523–529

Schoener TW (1968) The anolis lizards of Bimini: resource partitioning in a complex fauna. Ecology 49(4):704–726

Schmidberger M, Morgan M, Eddelbuettel D, Yu H, Tierney L, Mansmann U (2009) State of the art in parallel computing with R. J Stat Softw 31(1):1–27

Scutari M (2010) Learning Bayesian networks with the bnlearn R package. J Stat Softw 35(3):1–22

Scutari M (2012) bnlearn: Bayesian network structure learning, parameter learning and inference. R package version 3.2

Scutari M, Brogini A (2012) Bayesian network structure learning with permutation tests. Commun Stat Theory Methods 41(16–17):3233–3243 (special Issue "Statistics for Complex Problems: Permutation Testing Methods and Related Topics". Proceedings of the Conference "Statistics for Complex Problems: the Multivariate Permutation Approach and Related Topics", Padova, June 14–15, 2010)

Scutari M, Nagarajan R (2012) On identifying significant edges in graphical models. Artificial intelligence in medicine special issue containing the proceedings of the workshop "Probabilistic problem solving in biomedicine" of the 13th Artificial Intelligence in Medicine (AIME) conference, Bled (Slovenia), July 2 (in print)

Shäfer J, Strimmer K (2005) A shrinkage approach to large-scale covariance matrix estimation and implications for functional genomics. Stat Appl Genet Mol Biol 4:32

Smith SM, Fulton DC, Chia T, Thorneycroft D, Chapple A, Dunstan H, Hylton C, Zeeman SC, Smith AM (2004) Diurnal changes in the transcriptome encoding enzymes of starch metabolism provide evidence for both transcriptional and posttranscriptional regulation of starch metabolism in Arabidopsis leaves. Plant Physiol 136(1):2687–2699

Spector P (2009) Data manipulation with R. Springer, New York

Spirtes P, Glymour C, Scheines R (2001) Causation, prediction, and search, 2nd edn. MIT Press, Cambridge

Stein C (1956) Inadmissibility of the usual estimator for the mean of a multivariate distribution. In: Neyman J (ed) Proceedings of the 3rd Berkeley symposium on mathematical statistics and probability, pp 197–206

Sugimoto N, Iba H (2004) Inference of gene regulatory networks by means of dynamic differential Bayesian networks and nonparametric regression. Genome Inform 15(2):121–130

Tibshirani R (1996) Regression shrinkage and selection via the Lasso. J R Stat Soc B 58(1):267–288

Tierney L, Rossini AJ, Li NM, Sevcikova H (2008) snow: simple network of workstations. R package version 0.3-3

Tsamardinos I, Aliferis CF, Statnikov A (2003) Algorithms for large scale Markov blanket discovery. In: Proceedings of the 16th international Florida artificial intelligence research society conference, AAAI Press, pp 376–381

Tsamardinos I, Brown LE, Aliferis CF (2006) The max-min hill-climbing Bayesian network structure learning algorithm. Machine Learn 65(1):31–78

Venables WN, Ripley BD (2000) S programming. Springer

Venables WN, Ripley BD (2002) Modern applied statistics with S, 4th edn. Springer

Verma TS, Pearl J (1991) Equivalence and synthesis of causal models. Uncertain Artif Intell 6: 255–268

Whittaker J (1990) Graphical models in applied multivariate statistics. Wiley

Wu FX, Zhang WJ, Kusalik AJ (2004) Modeling gene expression from microarray expression data with state-space equations. In: Proceedings of the 9th Pacific Symposium on Biocomputing, pp 581–592

Yaramakala S, Margaritis D (2005) Speculative Markov blanket discovery for optimal feature selection. In: ICDM '05: Proceedings of the 5th IEEE international conference on data mining, IEEE Computer Society, pp 809–812

Yu H (2010) Rmpi: Interface (Wrapper) to MPI (Message-Passing Interface). R package version 0.5-8

Zhang H (2004) The optimality of naive bayes. In: Proceedings of the 17th International Florida Artificial Intelligence Research Society Conference, AAAI Press, pp 562–567

Zou M, Conzen SD (2005) A new dynamic Bayesian network (DBN) approach for identifying gene regulatory networks from time course microarray data. Bioinformatics 21(1):71–79

# Index

**A**
arc, 1
  directed, 1
  undirected, 1

**B**
Bayesian network
  causal, 44–46
  completed partially directed acyclic
    (CPDAG), 15
  continuous, *see* Gaussian
  discrete, 20–22, 42–43
  dynamic, 63–67
    homogeneous, 66
    non-homogeneous, 69
    VAR, 66–67
  equivalence class, 15
  Gaussian, 20–23, 40–42
  model averaging, 48
  moral graph, 16
  skeleton, 2, 18
  static, 13
boostrap, 48–51, 115–117

**C**
conditional independence test
  Fisher's exact test, 21
  Fisher's Z, 22
  log-likelihood ratio $G^2$, *see* mutual
    information
  mutual information, 21
  Pearson's correlation, 22
  Pearson $X^2$, 21
  permutation, 21, 22
  shrinkage, 21, 22

cross-validation, 117–120
cycle, 3

**D**
d-separation, 14–15
data
  `hailfinder`, 110–114, 116–122
  `isachs`, 53–56, 91–94
  `lizards`, 5–10
  `marks`, 26–46, 106–108
  `sachs`, 47–53
discretization, 23–24, 42–44, 47–48

**E**
edge, *see* arc, undirected
equivalence class, 30–31
evidence, 86
  hard, 86
  soft, 86

**F**
fundamental connections
  convergent, 15
  divergent, 15
  serial, 15

**G**
graph, 1
  acyclic, 3
  directed, 1
  ordering, 3
  partially directed, 1
  undirected, 1

R. Nagarajan et al., *Bayesian Networks in R: with Applications in Systems Biology*,
Use R! 48, DOI 10.1007/978-1-4614-6446-4,
© Springer Science+Business Media New York 2013

**I**
inference, 85–87
  approximate, 89, 93–94, 120–122
  belief updating, 85
  causal, 90
  exact, 87–89, 91–93
intervention
  inference, 90
  structure learning, 53–56

**L**
learning, 17
  parameter, 17, 23
    Bayesian posterior, 23, 91
    maximum likelihood, 23, 40, 42
  structure, 17–23
    constraint-based, 17
    hybrid, 20
    score-based, 19
    search-and-score, *see* score-based

**M**
map
  dependence, 13
  faithful, 13
  independence, 13
  isomorphic, 13
  perfect, 13
Markov blanket, 16

**N**
network scores
  Bayesian Dirichlet equivalent (BDe), 22
  Bayesian Gaussian equivalent (BGe), 23
  Bayesian Information Criterion (BIC), 22
  Minimum Description Length (MDL), 22
node, 1
  ancestor, 3
  child, 3
  descendant, 3
  neighbor, 3
  parent, 3
node ordering, *see* graph, ordering

**P**
path, 3

**Q**
query, 86–87

conditional probability (CPQ), 86–87,
  92–94, 120–122
maximum a posteriori (MAP), 87, 93
most probable explanation, *see* maximum a
  posteriori (MAP)

**R**
R, 4
  base distribution, 4
  contributed packages, 4
  CRAN, 4
R packages, 24
  **ARTIVA**, 80–81
  **bnlearn**, 24, 26–31, 34–43, 45–56, 91,
    93–94, 96–100, 110–114, 116–122
  **catnet**, 25, 33–34, 43–44, 50–51
  **deal**, 25, 32–34, 39–40
  **G1DBN**, 79–80
  **GeneNet**, 78–79
  **glmnet**, 74
  **gRain**, 25, 91–93
  **lars**, 74–77
  **pcalg**, 25, 34, 37
  **penalized**, 74, 96–100
  **Rgraphviz**, 34
  **simone**, 77–78
  **snow**, 105–108
  **vars**, 72–74

**S**
structure learning algorithm, *see also* learning
  structure
  Inductive Causation (IC), 17
  Auto-Regressive TIme VArying (ARTIVA),
    69–72, 80–81
  Fast Incremental Association (Fast-IAMB),
    18
  G1DBN, 68, 79–80
  Grow Shrink (GS), 18, 35, 42
  hill-climbing, 19, 37–40, 42, 45–46, 48–50
  Incremental Association (IAMB), 18, 35
  Interleaved Incremental Association
    (Inter-IAMB), 18
  James-Stein shrinkage, *see also* conditional
    independence test, shrinkage, 68, 78–79
  LASSO, 67–68, 74–77
  Max-Min hill-climbing (MMHC), 20
  PC, 18, 37
  simulated annealing, 19, 43–44, 50–51
  Sparse Candidate (SC), 20
  Statistical Inference for MOdular NEtworks
    (SiMONe), 69, 77–78
  tabu search, 19, 54–56

**T**
time series
  lag, 60
  multivariate, 60–61
  order, *see* lag
  stationary, 60
  univariate, 59–60
  vector autoregressive (VAR), 60–63
    lag, 62, 73

  normality, 62–63, 73
  serial correlation, 63, 73–74
  stationary, 61–62, 73
topological ordering, *see* graph, ordering

**V**
v-structure, 15, 16, 30–31
vertex, *see* node